Keep Your Own Livestock

A Practical Guide to Self-Sufficiency

CHARLES TREVISICK, F.Z.S.

Stanley Paul, London

Stanley Paul & Co Ltd
3 Fitzroy Square, London W1

An imprint of the Hutchinson Publishing Group

London Melbourne Sydney Auckland
Wellington Johannesburg and agencies
throughout the world

First published 1978
© Charles Trevisick 1978
Drawings © Stanley Paul & Co Ltd 1978

Set in Monotype Bembo
Printed in Great Britain by
The Anchor Press Ltd, and bound by
Wm Brendon & Son Ltd, both of
Tiptree, Essex

ISBN 0 09 129460 6 (cased)
      0 09 129461 4 (paper)

KEEP YOUR OWN LIVESTOCK
A Practical Guide to Self-Sufficiency

Contents

Acknowledgements

My grateful thanks are due to Jimmy Dillon White for all his invaluable help in the preparation of this book. Thanks also go to David Papworth for the line drawings and to Frank Lane, Sally Anne Thompson, *Poultry World*, The National Pig Breeders Association and Foto Powell Limited for allowing me to use photographs of which they own the copyright.

Foreword

The object of this book is to help the small, non-professional farmer to live off the land. Other books have been written about growing fruit and vegetables in your garden, about making wine and bread or, if you are artistically inclined, pottery.

This book is about animals.

In these harsh economic times any exercise in self-help makes sense: if you have land it pays you to use it.

Ideally you would wish for a small-holding of two to three acres (about a hectare), with a cottage or an old mill, a duck pond or stream and a range of outbuildings and perhaps access to some common grazing, but like any other exercise in self-help it is possible to make the best of what you have. A garden of half, or even a quarter acre can support a pig, a goat and a few rabbits and hens. This book tells you how to go about it.

I   Introductory

Thirty years ago, when I was a young man and had my prize herd of Friesians, my pigs and sheep and poultry down at Little Comfort in Devon, a farmer was a man who owned or rented two or three hundred acres or more. He was a man of substance, or comparatively so, with his farm and outbuildings, his livestock and land and expensive machinery. Today, commercial farming, in my opinion, requires a capital investment of £100000 or more. This is big money, and, let's be honest, the return is small when you remember the long hours, the hard, often back-breaking work and the risks.

So let us be clear about one thing: this book is *not* about commercial farming and certainly not about making money. It is about self-sufficiency.

In Cornwall, which is the next county to mine, there were and still are any number of part-time farmers. In the old days they worked half a day in the tin mines and spent the rest of their time on small-holdings. Now, with most of the mines closed, they have other jobs but they still have their livestock.

On the Continent – and remember that we are members of the EEC – two thirds of the farmers have less than 25 acres (10 hectares), and many have only moderately sized small-holdings, yet they all seem to manage. If, after reading this book, you think of becoming a backyard farmer you could do worse than take a trip abroad and see how our EEC friends make the most of their land.

But, of course, there are plenty of examples in this country. The two world wars threw up thousands of would-be part-timers, returning Tommies, appalled by the thought of dead-end jobs and crowded city streets, who looked to the land. They sank their gratuities and savings in small plots, often with dilapidated cottages or sheds as the only living quarters, bought up some animals and poultry and started to farm. They became small-holders or, if they could afford 30 to 40 acres (12 to 16 hectares), what we call 'wheelbarrow farmers'. Many failed, many returned, disillusioned, to city streets: many more stayed.

It wasn't easy, nor will it be easy for you if you decide to follow their example. In the first place there are never enough hours in the day; there will be set-backs and disappointments, sometimes frustrations and delays: any experiment in farming means long hours and hard work: it requires courage. And if, at the end of the day, some wiseacre points out that you could have done as well, or better, by investing in gilt-edged, you may wonder if it was all worth while.

But I doubt it. As one who has been fortunate enough to see a good deal of life and who has been in his time farmer, zoo keeper, author, television personality, animal breeder, prison visitor and Justice of the Peace, I can tell you that there is no satisfaction like farming, nothing to compare with that quiet moment at the end of the day when, with muddied boots, you lean on a gate post, watching pigs rooting in the paddock, the cows chewing thoughtfully in the byre and the hens settling for the night.

'You must be mad!' your friends will say when you tell them that you are going to buy a small-holding or even if you decide to keep a few hens and rabbits and perhaps a pig in the backyard.

Don't you believe it. If you appreciate what you are taking on, especially the work involved, you will get immense satisfaction from your venture, not to mention peace of mind. You can have meat and eggs for your table, butter, cheese, honey and fish: you can be virtually self-sufficient. With prices rising almost daily, it makes sense.

For inflation, it seems, is here to stay. £2·50 for a leg of lamb this week, £3·50 next: who knows what we may be paying a year hence? With the pound struggling like a feeble swimmer from one crisis to the next, it is hard to know, or even guess, where our money is going. If we work hard, if wages keep within sight of prices, if the oil-producers decide that enough is enough, we might – just might – maintain a reasonable standard of living. At worst, one feels, we should not go hungry.

But meat and poultry for our Sunday table, butter, eggs and honey for our tea? Only an optimist would bet on it.

Unless he is a farmer. I repeat: backyard farming makes sense.

I am going to assume that most of you who read this book will be amateurs or, at the best, part-timers, and that, whether you intend to buy a small-holding or to make-do with your present garden plot one of you, at least, will continue his present job. This is important, because while it clearly restricts the scope of operations it does mean that come what may you will have something to fall back on. There *are* such things as fowl pest and swine vesicular disease, as I shall explain in later chapters, and it is as well to have a second string to your bow.

Now, after weighing the pros and cons and examining the bank balance you decide that you want to have a go. Your present garden is too small or perhaps unsuitable for other reasons so you decide to move.

Ideally, of course, you wish for a small-holding of about 3 acres (1·2 hectares), with a cottage and perhaps some out-buildings and a stream which can be tapped.

Surprisingly, such properties are not all that difficult to find, especially if the ground has been neglected and the cottage needs modernizing or is in a poor state of repair. Prices will vary in different parts of the country and, so far as the 'second string' job permits, it is advisable to look in some of the less fashionable areas, for while the company director with his swimming pool and elegantly modernized cottage is a pleasant enough neighbour he does tend to keep prices up.

Find something near a market town if you can and choose another small-holder or working farmer for a neighbour. They may not have swimming pools or Jaguar XJ6s but they will talk your language.

When you find a place that is within your means don't be too dismayed if it looks a bit of a mess. Have a surveyor look over it, of course – this is essential – but if the building is structurally sound and has a roof which keeps out the rain you can move in quite quickly, despite the drawbacks, and you will be surprised at what can be done with a broom and a mop and a few tins of paint.

Start on the kitchen, I suggest, and then the bedroom, for discomfort is so much easier to bear if you can eat and sleep in comfort. Cold, dark corridors and draughty loos are all right for a week or two and can even be fun so long as you can laugh at them from the cosiness of a kitchen range or the warmth of a bed. It may take days or weeks or even months to make your cottage habitable, but until it is you should resist if you can the dappled sunlight in the paddock, the hum of bees in the orchard and the smell of honeysuckle in the hedgerow. First things first is not a bad rule of life: for the farmer it is a necessity.

Now one other point before I forget: to make your backyard farm a real success you will need a deep freeze. It's not essential, of course, but if you can possibly afford one you will find it money well spent. Even in the first year it should pay for itself; after that everything will be pure profit. Chickens, ducks, geese, all the succulent parts of a pig can be stored there: with a deep freeze you will never have to ask yourself, 'What shall we have for dinner?'

To return to the land: three acres (about one hectare) we said. Well, that is a nice manageable plot. Not for you the daunting sight of vast expanses of willow herb, of dock and nettle and the ubiquitous cow parsley. Even if your plot has really gone to seed you can work wonders in a few hours with a strong arm and a sickle or scythe (a scythe is quicker, once you have the hang of it – the point *will* keep sticking in the ground.) But it will be hard work. You will be hot and thirsty, your back will ache, your hands will blister, and the sweat will run continually into your eyes, but, as you work, think of all the bedding you are providing and the pleasure and comfort you will be giving to those birds and animals who will soon be sharing your land.

Of course, if you are less energetic you can always use a motor scythe or rotary grass cutter. Alternatively, there are excellent chemical sprays such as MCPA, which you can use with a knapsack sprayer. This is an ideal way to rid your land of broad leaf weeds such as nettle, docks and thistle.

If, like me, you have a reasonably tidy mind you will see the sense of dividing the plot into fenced strips. This is essential, for however attractive it may be to think of a cow, some pigs and goats and sheep, not to mention fifty to sixty hens, gathering to welcome you at the kitchen door you have to remember that goats and sheep are heavy, pigs more so, while cows are positive heavyweights. A grazing animal allowed to wander without restriction will spoil more grass than it will eat, so the sensible farmer rations his pasture, as we must, with the help of fences.

How many fences? Well, with three acres I suggest that you divide your land into six half-acre (0·2-hectare) plots. Four of these will be for the grazing animals, one for the poultry, while the sixth will be used as arable land to grow root crops for stock and to serve as a kitchen garden. In fact, provided your soil is not marshland or desert you should have no difficulty in becoming self-sufficient in vegetables, while if you are lucky enough to inherit one or two apple trees – and few country cottages are without their Worcester Pearmain or Grannie Smith – you can avoid most of those expensive trips to the fruiterers. This book, as we said before, is about animals, but animals and vegetables go together, especially on a smallholding, and both are important to any exercise in self-sufficiency.

Divide your land, I said: but how? Wire mesh cattle fences 3 ft 6 ins (1 m) high are ideal but can be expensive, so if you are pushed for cash and the bank manager is already giving you old-fashioned looks you can do equally well with single-strand fences carrying a mild electric current.

This paddock system, which is used by most dairy farmers, is wonderfully economic, and your plot, so divided, will support up to twice as many livestock as would have been possible if your cow, your goats and sheep and pigs had been allowed to roam free.

Of course, the poultry plot in which you will keep your ducks and hens and other domestic birds will need to be fenced more securely, both to stop the birds from straying and to keep out marauding dogs and foxes. From experience I recommend 2 in (5 cm) netting, 6 ft (1·8 m) high, supported by larch poles and buried at the base 6 ins (15 cm) into the soil. This takes time, time you feel you can ill afford, but it will be worth it. In all my years of poultry keeping I have never lost a bird to a burrowing stoat or fox.

Well now, with your cottage habitable, although perhaps still not to the standard of comfort and decoration you intend, your land cleared of undergrowth and your fences in position you can think of livestock.

You have a wide range to choose from, perhaps wider than you think and it will now be your pleasurable duty to decide which animals to buy.

In this book we shall discuss the livestock that can reasonably be kept on a small-holding or large garden, but it should not be inferred from this that all or most of the birds and animals described would necessarily be suitable for your plot. In each chapter I have tried to indicate what you are likely to need, both by way of food and housing, and I have mentioned the relative risk of disease.

Before deciding on any animal you would be well advised to prepare a costing, based on current prices, to prove that if the animal or bird is kept and fed for so many months you will not only be able to cover your expenses but should make a profit. Pigs, for instance, especially if kept in small numbers, are a doubtful investment, although this may change, and in any case the baconer fattened for your own table will always be worth while.

Intelligent costing will help you to choose widely, experience will prove whether you were right. There is no magic in farming, only common sense and hard work.

So, what can we keep?

In this book we shall discuss breeding and keeping hens and ducks for eggs and table meat.

We shall consider turkeys, guinea fowl and geese.

We shall consider rabbits, which are easy to breed and make good eating, and pigs.

We shall discuss goats, which will thrive almost anywhere, and, with the minimum of attention, will provide milk and table meat.

We shall take a look at sheep, which, even for a small-holder, can be profitably bred for wool and meat.

We shall consider a milking cow, almost essential in any scheme for self-sufficiency because of its milk, butter and cheese.

We shall consider a hive of bees.

We shall discuss tapping the stream that runs through your plot, tapping it in two places perhaps, one for the ducks in their enclosure, one for a stewpond of fish.

Meat and eggs, butter, cheese and fish: all these and more could come from your own plot. All you need is courage, patience and a capacity for work, and – let's face it – a measure of luck. It won't be easy, there will be disappointments, but at the end of the day you will be able to look back and remember, as details of a bad dream, the ever changing labels in the supermarket and the skewered tags at the butcher's. You will be able to grin, as I do, and say, 'Inflation? So what?'

2 Domestic fowl

It came to me out of the blue when I was least expecting it. 'Which is the most common bird in the world?'

My mind, anticipating questions on the endemic diseases of guinea fowl or the feeding processes of the pelican, went blank. The other members of the panel looked at me impatiently and thrummed the table – after all, wasn't I the bird expert?

Most common? The sparrow, I thought. There must be millions of them in this country alone. Or the finch, which is just as familiar and is found in varying shapes and sizes throughout the world? No, it must be something more exotic; the parrot, perhaps, or one of the sea birds.

Then the penny dropped. I had not been a member of the Poultry Club all those years for nothing.

'The domestic fowl,' I replied, and took a sip of water, knowing I was right.

Yes, your common or garden chicken is the most popular bird in the world, and with good reason, for whether your backyard is in Glasgow or Seattle, Kingston, Jamaica or Timbuktu you need only set her down with a patch of earth to scratch, some scraps to eat and a corner to nest in and she will do you proud.

She will provide eggs for your table, chicks to rear or sell and, when she comes to the end of the day, meat for your table and feathers for your bed. She is easy to look after, no more prone to illness than any other domestic bird, and indeed if she is given clean quarters and a dust bath to shake herself in she is virtually trouble-free. What a paragon of a bird!

HOW TO START

Nevertheless, as a small-time farmer your poultry will be an investment, one of your most profitable as like as not, and as with any other investment there are points to watch.

Although, or perhaps because, you are new to the game, I advise you not to go rushing to buy the first hens on offer. In the local paper or perhaps at the market you will find birds that are cheap and apparently healthy, but which may well prove to be indifferent layers or, if they are cockerels, will remain scrawny and nondescript however much you feed them.

Go to a reliable poultry farm and buy eight-week-old pullets or, if you can obtain them, some 'point of lay' pullets.

The most popular and trouble-free breeds in my opinion are Rhode Island Reds and Light Sussex – I can recommend both but you may have difficulty in obtaining stock. Over the past twenty or so years, the poultry industry has changed from small-scale farm units to large commercial enterprises, run by business companies.

These companies now breed 'hybrid' birds which are a mixture of the pure breeds. Hybrids grow quickly and produce more eggs than our native pure breeds, but do require special attention if they are to give the best results.

You can often purchase hybrid hens after one season's laying, directly from farmers. These birds, which have been kept in battery cages will quickly respond to your outdoor run and lay 'free range' eggs in abundance as they enjoy their new-found freedom.

Now before you buy your pullets you must decide where and how you are going to keep them.

FREE RANGE

The ideal system, of course, but not, alas, for you, is what we call free range. You could once see this on general farms where the chickens were kept more or less as a side line and allowed to roost in farmyard sheds or barns or, in substantial houses on

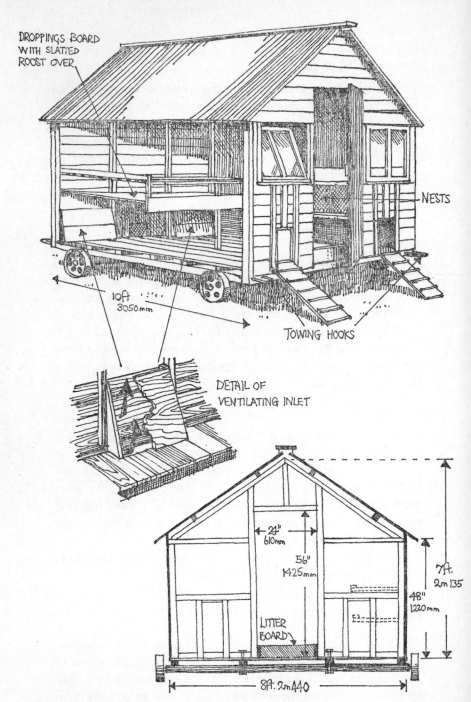

DROPPINGS BOARD
WITH SLATTED
ROOST OVER

NESTS

10ft
3050mm

TOWING HOOKS

DETAIL OF
VENTILATING INLET

24"
610mm

56"
1425mm

7ft.
2m 135

48"
1220mm

LITTER
BOARD

8ft. 2m 440

Portable poultry house

wheels which could be moved from field to field. The chickens were allowed to run free and, apart from the saving in food – for they found most of what they needed among the hedgerows and stubble – they remained healthy and productive.

But with your limited space we must think of other methods.

THE BACKYARD FOWL HOUSE

Let us consider first of all those readers who only have backyards of, say, a quarter of an acre (0·1 hectare) or less. They will have to think in terms of a permanent fowl house set in a sunny corner of the garden or, perhaps, against a wall of the house.

There are various designs, which you can sometimes see at corn chandlers or pet shops, but the points to remember, whether you are buying or making your own, are as follows.

Dry Floor

A dry floor is essential, so, if possible, set your house on balks of timber or, better still, on bricks.

Perches

Fowls in the natural state roost in trees, so yours will need perches, set at least 2 ft (60 cm) from the ground. To give maximum comfort the perches should be 2 ins (5 cm) wide and not less than 10 ins (25·5 cm) from the wall. Allow 8 ins (20 cm) of perch space for each bird, and if you have two or more perches see that they are at least 14 ins (35·5 cm) apart.

Dropping board

To keep your birds healthy you must ensure that their house is clean, so a dropping board placed 6 ins (15 cm) or so below the perches is essential. Make sure that you can get at it easily, for it will need to be cleaned daily.

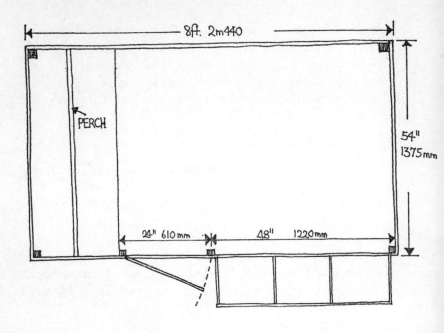

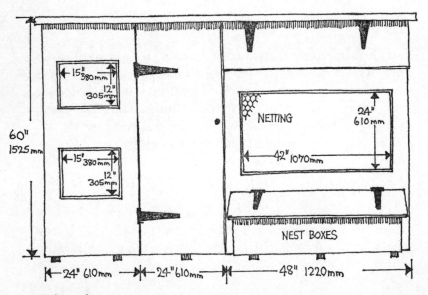

Simple poultry pen

Nests

Now for the most important part, the nests. These should be away from the perches, otherwise the birds may roost in them, and against an outside wall so that you can easily remove the eggs. Each nest of straw or hay should be set in a compartment, approximately 2 ft (60 cm) square, with a base of wire netting to keep out the rats. You can build a colony of nests, three or four or more in a row, but remember that each should be enclosed on three sides and you should allow one nest for every three birds.

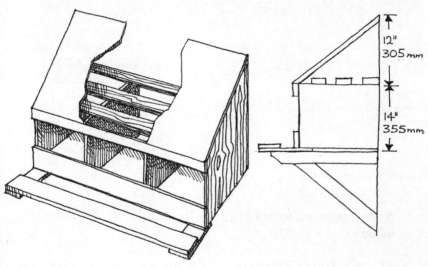

12"
305 mm

14"
355 mm

Nest boxes

Artificial lighting

If you want to increase your egg yield, especially in winter when prices are high, it will pay you to rig up some simple form of lighting. A low-watt bulb run from a battery or from the electricity supply in your own house will give extra 'daylight' or feeding time, and your hens will repay you with more eggs.

Now, having bought or built your fowl house there are

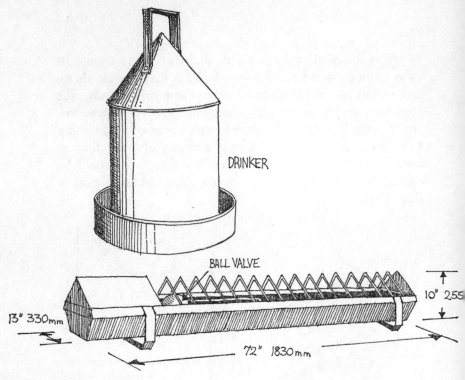

DRINKER

BALL VALVE

13" 330mm

10" 255

72" 1830mm

Drinker and self-filling water trough

certain essential furnishings which can be summed up as follows.

Fresh water supply

Your hens will need a constant supply of fresh water, so I suggest a galvanized iron water fountain, which is self filling, and that you allow 1 gallon (4·5 l) of water for every six birds.

Food troughs

To make sure that they all get a fair share of food you should buy, or construct (they are quite simple), V-shaped food troughs, allowing about 2 ft (60 cm) of trough for every six fowls.

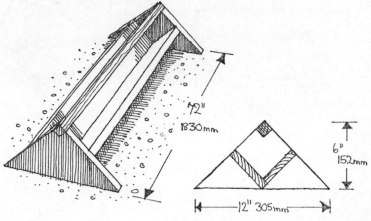

Simple feed trough

Dry mash hopper

Unfortunately, although the food trough is suitable for wet mash or greenstuff it is not suitable for some other types of food. If you put in dry mash or pellets, for instance, your hens will hook over the side as much as they eat, which is wasteful and unnecessary, so you can make a simple hopper (as opposed to the much more elaborate and expensive ones you can buy in shops) with a long, open topped wooden box with a narrow batten set longways about 1 in (2·5 cm) from the front and a similar distance from the top.

Grit trough

Grit and shell are almost as important to your hens as water. Flint or gravel is necessary for their digestion, limestone grit or oyster shell will supply the lime for their eggs. Both types of grit and fish shell should be provided in a simple form of double compartment grit trough. If you only have a few hens you can make a self-filling box, as in the diagram overleaf, which will serve for either grit or dry mash. In fact the handyman will think of all sorts of gadgets which will be both cheap and effective.

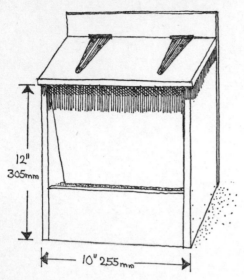
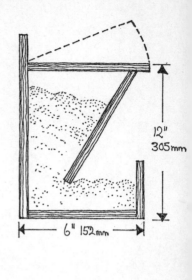

Self-filling grit and dry mash hopper

Dust box

It seems strange that birds should keep themselves clean by wallowing in dust, but this is nature's way, and it works. Your chickens will almost certainly be troubled by small insects and parasites unless you give them a box filled with dry earth, mixed perhaps with ash from your fire and a little sulphur powder, in which they can 'bathe'. It is a natural process for them, of course, so they won't need teaching, but they do tend to make rather a mess, so you may prefer to put the boxes in the fowl house once or twice a week.

Litter

Before you put your pullets in their new house you should cover the floor to a depth of a few inches with dry litter in which they can scratch. Straw or peat or wood shavings are all suitable, if possible laid on a base of sawdust or dry ash.

POULTRY FOLDS

Now, having considered the reader who is limited to a back-yard, let us consider the more fortunate one who has a small-holding with perhaps half an acre (0·2-hectare) wired off for poultry.

Instead of the permanent fowl house I would recommend for him folds, which are simply portable houses with wire netting runs attached. They are quite simple to construct and each should be large enough to accommodate about twenty birds.

The folds are moved each day or every other day so that the hens, or pullets in your case, can pick up insects and grubs as well as their favourite diet of short, sweet grass, while, as an added bonus, they will be dropping valuable manure on your land. The folds will protect the birds from dogs or foxes and the eggs from hawks and carrion crows.

However, it is only fair to say that fold units are heavy to move, especially in winter when the land is wet and sticky.

BASIC CARE OF YOUR POULTRY

Feeding

Feeding your pullets is simple and relatively inexpensive. From their run, especially if the folds are moved daily, they will find insects, seeds, grubs and fresh grass. To supplement this they need two meals a day in summer, three in winter, of poultry meal mixed with boiled scraps of bread, potato and plate scrapings. On a cold winter's morning I find they appreciate a mixture of scalded flaked maize and kitchen scraps, given while still warm. For a change I throw handfuls of mixed grain – wheat, barley, oats and maize – into the litter where they gain healthy exercise scratching for it.

You can reckon that a fully grown hen eats about 4 ozs (100 g) of food a day of which 2 ozs (50 g) – a good handful – should be grain. If your hens are kept in a permanent house rather than a fold you will have to supplement their diet with greenstuff such as cabbage leaves or handfuls of grass.

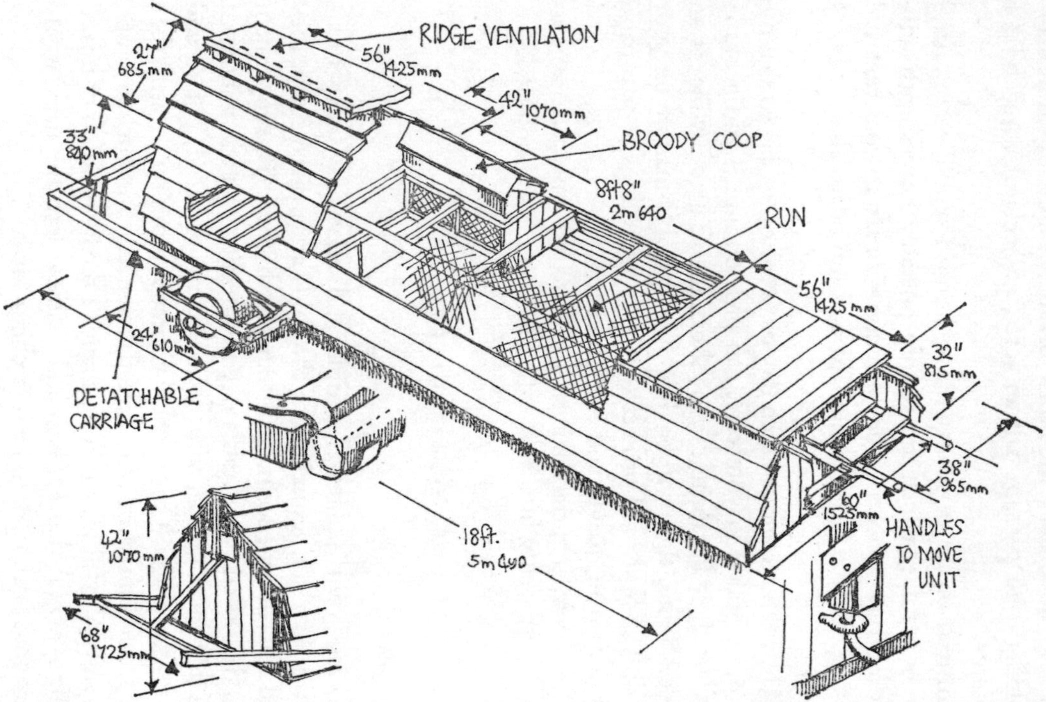

RIDGE VENTILATION

56"
1425mm

42"
1070mm

27"
685mm

33"
800mm

24"
610mm

BROODY COOP

RUN

8'8"
2m 640

56"
1425mm

32"
815mm

38"
965mm

60"
1525mm

HANDLES
TO MOVE
UNIT

18ft.
5m 490

DETATCHABLE
CARRIAGE

42"
1070mm

68"
1725mm

Portable poultry fold unit

And, whatever you do, don't forget the 'musts' we discussed earlier – the supply of fresh water, the two kinds of grit and the powdered shell.

Egg laying

Now to the most important item of all, the egg laying.

Your pullets should come into lay when they are about six months old (a week or two before if they are in very good condition), they will continue at their best for another twelve months, after which their production will fall year by year. The useful laying span of most birds is therefore only about two years; the older the bird the less she will lay, especially in the winter months when egg prices are highest, so it is usual to kill off all but a few specially good hens needed for breeding after the end of the second laying year. And in case any of my readers feel that this is a bit ruthless and that besides they could never bring themselves to do it, remember that this is a commercial venture like any other and your aim should be a small farm unit run profitably as well as efficiently. As for the actual killing you can, if you are squeamish, get your local butcher to give you a hand. He will do it quickly and humanely.

If you ask me how many eggs you can expect I would say that it depends on a number of factors, the breed of bird you have chosen, the health and age of your hens and whether you can give them artificial winter lighting. You will probably hear many tales of prodigious output, especially from your more experienced neighbours, but remember that these are rather like anglers' tales, if not wilful exaggerations then at least to be taken with a grain of salt.

Your pullets, starting in, say, October, may produce two or three eggs a week to the end of December, increasing to four or five a week in the following March or April. This is without artificial lighting. If you give them an extra two or three hours' feeding time in the winter months they could easily increase this during that time.

Artificial lighting

This helps to increase winter egg production in two ways. Firstly the light stimulates the ovary via the pituitary gland and so more eggs are laid.

Secondly, by extending the feeding time to around 14 hours per day, the birds are able to eat more food and consequently will lay larger eggs.

The usual practice is to install a time switch in the electric circuit. This can be set to switch the lights on early in the morning and off when there is sufficient daylight for the birds to feed.

In the early spring when daylight extends to 14 hours the lighting can be dispensed with.

Normally a 100 watt bulb hung over the feed troughs is sufficient. Allow one light per 20 ft (6 m) run.

In the second year, when they are fully grown hens, production will fall, but not by much, and, as I said before, it is not until the third laying year that their egg production really falls.

So, to sum up, with a good bird and artificial lighting you might expect 200 to 250 eggs in a full year: top grade eggs weigh 2 to $2\frac{1}{4}$ ozs (50 to 55 g) each and second grade eggs $1\frac{3}{4}$ to 2 ozs (45 to 50 g). By any account that is a lot of eggs.

The moult

Don't be dismayed if, or rather when, egg production drops in late summer or early autumn. Chickens change their feathers once a year, and while this moult, as it is called, is on egg laying drops and frequently ceases. The moult, which lasts about two months, usually occurs in July or August, but for a good laying hen it is more likely to be September. Most poultry farmers kill off their older or less productive hens as soon as the moult starts and before the birds lose condition.

BREEDING AND REARING

Now let us assume that you have got through your first

summer and winter, your pullets are laying nicely and, full of confidence, you start thinking of expansion.

One way, of course, would be to buy more pullets, but the ones you bought last year are just coming into their prime, none has been such a poor producer that she deserves to be culled (killed to the layman – you will soon find yourself using farming language), and in any case you are only producing eggs: what about table meat?

So you need cockerels as well as pullets.

Day-old chicks

One way would be to go to a reliable poultry breeder and buy day-old chicks. Perhaps one or more of last year's pullets, now fully fledged hens, is broody or, failing this, you can buy or rig up an artificial brooder.

Consider first your broody hen. She will take quite naturally and kindly to her foster chicks so long as you introduce them to her gently. Let her sit on a nest of china eggs for a few days, then, on the evening when the chicks arrive take one and place it under her breast. If she takes to this, as she almost certainly will, wait an hour and then introduce the others.

All the housing you need will be a coop, roughly 2 ft by 2 ft (60 cm by 60 cm), with a slatted front so that the chicks, leaving the warmth and protection of their foster mother, can venture out into a run where they will scrap about and find the odd insect or two before rushing back to mum.

If none of your hens is broody you can do equally well with a small heated pen, or coop, known as a brooder. The chicks will still enjoy the freedom of the run but will quickly find where to run back to for comfort and warmth. A suspended 25v electric light bulb will provide warmth.

Incidentally, many poultry farmers prefer to place day-old chicks in a brooder rather than use the services of a hen. This saves time, which is an obvious advantage, and you should have no difficulty in selling off any surplus chicks at eight weeks.

Breeding – the natural way

The natural way, of course, is to mate some of your best hens with a cockerel and then have the fertile eggs hatched by a broody hen or an incubator.

On the other hand you may feel that it would be easier to buy a clutch of fertile eggs from a reliable poultry farmer and to concern yourself only with the hatching.

Timing   The first and most important point is the timing of the operation. The experienced poultry farmer sets his eggs from early February to the middle of April, so that the chicks are less likely to experience severe weather and, if things go well, any resulting pullets will be in lay well before the winter. (Birds for table meat, whether cockerels or pullets, can be hatched at any time during spring, summer or autumn.)

One snag, though, as you will find, is that unless you are an expert you will have no means of telling which of your day-old chicks are cockerels and which are pullets. Not until the chicks are from six to ten weeks old can you say with any certainty, for by then the cockerels will begin to show more colour in their combs and wattles, and their tails, which will be developing into balls of feathers, are different from the straighter tails of the pullets. But what an advantage it would be to know one from t'other at birth.

Well, there are two answers to this problem; you can either take your day-old chicks to a hatchery where, for a fee, experts will sex them, or you can rely on sex linkage.

Sex linkage   It has been found that in certain crosses between pure breeds of one colour with pure breeds of another the pullets, even at the day-old chick stage, will show the colouring characteristics of the father and the cockerels the colouring of the mother.

The most common sex linkage is between 'gold' cocks and 'silver' hens, thus if we mate a Rhode Island Red (gold) with a Light Sussex the pullet chicks will be brown, after the father, and the cockerel chicks creamy white, after the mother. It's as simple as that.

Hatching

The time is right, you have your broody hen, or hens, for there are advantages in setting two or three hens in the same nest house: now before buying the fertile eggs there are a few preliminaries.

The nest house Ideally you need a quiet shed where your 'broody' will sit with several other hens, each on her own nest, until the chicks are hatched. Each nest should be in a box, approximately 2 ft (60 cm) square, with an earth floor to keep the moisture in and with straw or hay on top to form a comfortable bed. Under the straw and above the earth should be a protective layer of wire netting to keep out any mice or rats.

The broody hen Make sure that your hens are really broody by testing them on a nest of china eggs, as many hens are sat before they are ready and leave the nests before the twenty-one days are up.

Another word of warning: don't, unless you are prepared to take a chance, sit a light breed hen like a Leghorn or Ancona, for, as any experienced breeder will tell you, they can be unreliable and flighty, and come off the nest before the three weeks are up prior to hatching.

Once your broody hen shows that she has settled replace the dummy eggs with fertile eggs – a baker's dozen will be all she can comfortably manage – and let her get on with the job. She will need a minimum of attention, although it is only common sense to make sure that she is kept as quiet as possible and is not disturbed by children or barking dogs.

She should be allowed, or enticed if necessary, off the eggs once a day for food and water, ten minutes a day for the first week, fifteen for the second, and twenty for the third. On the eighteenth and each succeeding day, particularly if the weather is dry, it will pay you to dampen nest and eggs with warm

B

water while she is feeding. (You only need to sprinkle the eggs, not drown them.) The chick has quite a tussle to peck its way into the outside world, and in dry weather the inside skin and shell tend to harden. There is nothing more discouraging after the hen has completed her sit than finding that the chick has given up the struggle and is what poultry farmers call 'dead in the shell'.

You will notice each day when the hen returns to the nest that she turns the eggs with claw and beak. This is nature's way of ensuring an even warmth and is what you must do, as I shall explain later, if you use an incubator.

On the nineteenth day the first chippings from the emergent chicks should begin, and by the morning of the twenty-second day the hatching should be complete. Don't be upset if you have a few failures: on average only about 70 to 80 per cent of a clutch will actually hatch.

After that the procedure is the same as we have described for any other day-old chicks. The hen sits in her coop watching over her lively brood while you, greatly heartened by these visible signs of success, plan your next move.

The incubator

If, after this first success, you feel that you would like to go in for hatching on a bigger scale, and remember that there is a profitable and ready market for any surplus chicks, you should invest in an incubator. You will find it quite simple to understand and manipulate and, if you shop around, not all that expensive.

From experience I would advise you not to buy an oil-heated machine but to get a modern electrically controlled model, which – power cuts excepted – is virtually fool-proof. Having said this, you would be wise to give it a trial run for a few days before buying your eggs to make sure that it keeps the steady temperature of 39·5° C (103° F) required.

The incubator, which basically is no more than an electrically heated box, should be set on a level floor and clear of draughts, although fresh air is necessary. Turn the eggs twice daily, morning and evening, and see that the water trays, which maintain the right level of moisture in the air, are replenished. When you purchase a new incubator you will get a booklet giving operating instructions.

You should also check daily that the correct temperature is maintained.

Keep a sharp look-out for vermin – this applies to all your poultry houses. If you find any signs of rodent droppings, place small quantities of rat bait on the overhead beams or directly into any rat holes you may find. Take special care to see that the bait is kept away from children or pet animals.

From the third to the seventh day it is advisable to cool the eggs for five minutes by leaving the incubator door open and to double the cooling period to ten minutes from the eighth to twentieth days. In some incubators the eggs do not need to be cooled off in this way, as the time taken to turn the eggs by hand suffices. This will be explained in the introduction booklet.

On the twenty-first day, if you have taken these few precautions, the shells will have cracked, tiny beaks will appear, and you will have the thrill of seeing more visible returns from your investment.

Foster mother or brooder

Of course newly born chicks, however lively they may appear (and they are chirpy little creatures after only a few hours) are too small to fend for themselves. They can pick up grubs and insects and small grain within a few days but not enough to maintain the necessary body heat.

So it is usual to introduce them to a broody hen, as we described earlier, or, if one is not available, to keep them in a brooder. Until they are able to forage for themselves they should be fed on some small chick mixture with chick crumbs, a balanced diet which can be bought from your corn merchant.

Growers

At about six weeks the hen goes off and leaves her chicks,

which is, no doubt, a relief to all for she is a fussy and demanding mother, and the chicks, now able to fend for themselves and officially classed as 'growers', can be transferred to adult folds. The pullets will be your future egg producers and at five to six months should begin to lay. The cockerels will fatten to eating birds and should be ready for the table at about the same time.

So, consider the cycle, which is short enough and marked by sufficient evidence of progress to convince even the most doubting of Thomases that this side at least of his farming venture will be profitable: first, the pullets, bought from a reliable local breeder. These will start laying at about six to seven months old and are known as 'point of lay' pullets.

Then the day-old chicks or the clutches to be hatched by broody hens or incubators. Allow three weeks for hatching and another five or six months for them to develop into layers or saleable table meat.

Once started, the cycle can be repeated *ad infinitum* and, provided you use folds rather than letting your birds run free, you can soon rear a hundred or more birds on your small plot.

HEALTH AND DISEASES

Only a few years ago any chapter on poultry would have included a daunting list of diseases, but now, thanks to antibiotics, these have either been eradicated or controlled to a minimal degree. This is good news for the backyard farmer.

Healthy birds are easily recognized by their alert appearance, bright combs and wattles and the distinct sheen on their plumage. The birds will usually be found scratching around or taking a dust bath if the weather is fine. Occasionally you find a bird that is unwell – she may be found just standing around taking little interest in her surroundings or fellow birds, or with her head tucked under her wing. In cold, wet weather birds often suffer from cramp. This is due to poor circulation arising from the damp conditions underfoot. Like humans they also suffer from the common cold and you may find them with runny eyes and sneezing. The remedy in both cases is to house

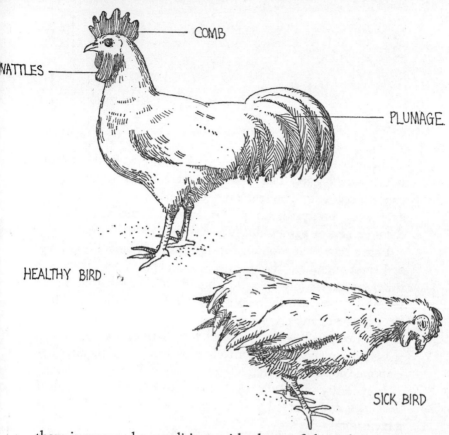

COMB

WATTLES

PLUMAGE.

HEALTHY BIRD

SICK BIRD

them in warm, dry conditions with plenty of clean, dry straw for bedding.

As with all livestock, good housing and adequate feeding is the best prevention of disease. You will lose the odd bird no doubt, but remember birds die from natural causes as well as disease! If, however, you should have the misfortune to lose several birds or you suspect that all is not well with your flock – a sudden drop in eggs or a change in normal behaviour – then telephone either the Ministry Veterinary Department or the Poultry Advisor of the Local Authority, for expert advice.

With proper care the risk of fowl disease is small. As I said at the beginning of this chapter, the domestic fowl is a paragon of a bird, hardy, productive and trouble-free. You will find out after you have been keeping them for a year or so why it is the most popular bird in the world.

3   Ducks

The chicken, we said, is the most popular bird in the world, but I would guess that if a poll were taken among those experienced with poultry, on the farm or small-holding or even in the backyard, a high proportion of votes would go to the common or garden duck.

I have kept these modest, nervous, likeable birds for as long as I can remember, first when I was a child as pets, then as a good profitable investment when I was farming at Little Comfort, then, as zoo keeper and breeder, for display. They are still one of my favourites.

For the small-holder or backyard farmer the duck has many advantages: it is easier to keep than a hen, it requires less elaborate housing and is much less prone to disease. It does not, as is generally supposed, need a pond to swim in, although if you are going to breed you will get much better results if you have natural water.

For the reader with a sizeable backyard a pen of ducks, especially if you are primarily interested in eggs, is quite feasible. All they will need are warm, dry sleeping quarters and a run of approximately 20 sq yds (17 sq m) a bird.

The sleeping quarters can be easily constructed, for the duck, which roosts on the ground, needs no perches, and in lieu of a pond you need only provide water troughs for the birds to 'duck' their heads in, or, if you want to do them proud, a galvanized iron preserving pan or hip bath sunk into the ground.

WHAT TO BUY

There are several well established domestic breeds, all with known characteristics and performances, and your choice need only depend on whether production of eggs or table meat is your aim.

All breeds, apart from the Muscovy, stem from the wild duck or mallard: the Muscovy, although a domestic breed, is the odd one out. It was introduced to this country from South America in the seventeenth century, and there is still some doubt, even among the experts, as to whether it is truly a duck or a goose. Since in this country it is kept mainly for show, although in Australia it is a popular table bird, I do not think it worth the attention of the backyard farmer.

Table birds

In this country most birds are kept for table meat since the white tender flesh is a favourite dish in homes and restaurants the year round, while the duckling, which is especially popular during the green pea season of early summer, is considered a delicacy.

The most popular table birds are Aylesbury, Pekin and Rouen, all of which are big fleshy birds and quick growers. A ten-week-old table duckling will weigh as much as 5 to 6 lbs (2·3 to 2·7 kg) – a chicken will take twenty weeks to reach the same weight – which means that the backyard farmer has a quick turnover and profit. Eight to ten weeks should see your ducklings ready for the table.

Of the three breeds my personal choice would be the Aylesbury, which is hardy and docile and a quick grower, and although it will lay very few eggs – perhaps thirty to sixty a year – its performance in the growth and profit stakes is unsurpassed.

Egg producing birds

Perhaps the main reason why ducks are less numerous than fowls on most farms is that duck eggs are something of an acquired taste, especially among town dwellers, although your true countryman will prefer a good large duck egg for his tea. The eggs are larger and richer than hen eggs, but do not keep as well, and perhaps it is this, rather than the strong taste, which makes them less popular in towns.

The best breeds for egg production are the Indian Runner, which was introduced to this country from Java in the last century, and the Khaki Campbell, the result of a cross between Mallard, Rouen and Indian Runner.

For you I strongly recommend the Khaki Campbell, which will cheerfully lay up to 300 eggs a year or more, is hardy and will clear your plot of slugs and snails and other plant devouring pests.

Unlike the chicken a duck lays most, or all, of its eggs in the early morning, which is difficult on a free range farm since they tend to lay in out-of-the-way places and cover their eggs with grass or straw. Unfortunately they may succeed in hiding them from the farmer but not from the fox or rat.

For you, however, this should present no problem, for if you shut your birds in at night you will not only protect them from foxes but you will find the eggs waiting for you in the morning providing you keep the ducks in until about 10 a.m. A clean, dry house with a raised floor and a bedding of straw or sawdust is ideal.

One word of warning: the duck egg shell is porous and, if laid on dirty ground, infection can seep into the egg, so it is doubly important to keep your duck house clean.

FEEDING

Ducks are cheaper to feed than hens, especially if they have a free run of your poultry plot, for they are great foragers and pick up slugs and worms and even snails if they can find them as well as insects and seeds.

I feed mine with a handful of grain for each duck before they are let out in the morning and, later, with a warm mash of kitchen waste and plate scrapings mixed with laying mash. In winter I increase this by adding warm flaked maize.

BREEDING

Now, having lived with ducks for a while and appreciating

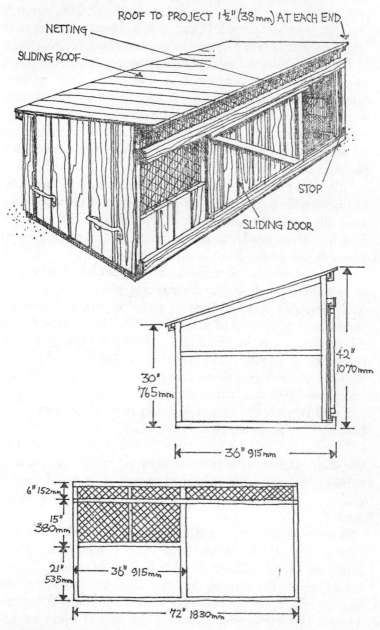

ROOF TO PROJECT 1½" (38mm) AT EACH END

NETTING

SLIDING ROOF

STOP

SLIDING DOOR

30" 765mm

42" 1070mm

36" 915mm

6" 152mm

15" 380mm

21" 535mm

36" 915mm

72" 1830mm

Duck house

their hardiness, their amiable disposition and, not least, their profitability, you decide to breed. Good! I don't think you will regret it, especially if by this time you have made your contacts with local butchers and hotels for the sale of surplus ducklings.

The first consideration, of course, is water. If you are going to breed from your own stock you will only do so success-fully, or shall we say profitably, if your ducks and drakes have access to water. Mating in the wild is done on pond or stream, and your birds, although domesticated, are not so far from nature that they can mate with any degree of certainty in other conditions.

Still, all is not lost. If you have no water on your land you can buy fertile eggs from a local breeder and, provided they are really fresh, that is, not more than six days old, you can introduce them gently, as we described in the last chapter, to a broody hen or put them in an incubator.

Nevertheless, natural breeding is best and for this, as well as a stream or pond, you need a breeding pen consisting of one really first-class drake and six or seven of your best ducks. I hope I can't be accused of male chauvinism if I suggest that in this entourage it is the drake that matters. Go for a good bird, for whatever it costs will be money well spent, and if your bank account is edging near the red you will do better with one really good bird than with two second-raters.

Similarly with the ducks: pick out those with good, deep bodies and in first-class condition.

The drake should be kept with his wives in a separate pen and with access to natural water until the fertile eggs are laid.

Now, perhaps when you least expect it, you will have a problem, for the duck, in the domestic state, is not an ideal mother.

In the wild or even on a free range she will settle down in a natural declivity among the reeds by a pond or stream and hatch her brood without trouble (unless, of course, rats or foxes discover her nest, as they probably will). Offer her the security of a ready-made nest in a shed and she will become nervous and introspective, leaving the eggs at every strange noise and, more often than not, will fail to stay the course.

So, you will need a foster mother, and for this you can't

improve on the broody hen. Choose a heavy breed such as the Light Sussex or a first cross between Light Sussex and Rhode Island Red. They are soft feathered birds and make good mothers.

The hatching procedure is very much as for hen's eggs except that the incubation period is twenty-eight days instead of twenty-one and the eggs will need more moisture. Sprinkle them every few days with warm water while the hen is off the nest.

It is also advisable to test the eggs from time to time during the incubation period to make sure that they are fertile. Hold them up to the light: the fertile eggs will show up much darker than the infertile ones, which are clear. The infertile ones should be removed.

On the twenty-eighth or twenty-ninth day the hatching will be complete and you will be rewarded by the sight of a dozen or so delightful ducklings.

REARING

Don't be too concerned about feeding them for a couple of days, for, like chicks, they will be sustained for at least this period by the embryo of the eggs they have just left.

After this and for the first week or more they should be fed on chick rearing mash, moistened to make it just crumbly, and chick crumbs. I suggest you give it to them in a dish or trough for, like most young creatures, they are mucky eaters.

Ducklings grow quickly and need a lot of food to maintain their growth. Cooked kitchen waste and plate scrapings help to fill them up, and they will take kindly to an additional diet of mashed potatoes. If you have a vegetable garden you can make a cheap additional feed by picking out the small potatoes and feeding them to your ducklings, cooked and mashed.

Before long they will start foraging for themselves and, given the run of the poultry plot, they will pick up slugs, snails and insects as well as green vegetable matter.

One word of warning: I said earlier that ducks are less prone to disease than hens, but they do have one surprising weakness: they are susceptible to sunstroke. 'Sprawls' we call it in my part of the country, and the person who looks after the poultry will try to keep young ducklings from strong sunlight.

They are also liable to chills if they emerge from wet, shady grass into the sunlight, but apart from these two hazards they are usually remarkably healthy and active.

SEXING

You will soon be able to tell ducks from drakes by their plumage. The drakes have brighter plumage and develop curled feathers in their tails in contrast to the more modest plumage of the duck.

The drakes and probably most of the ducks, unless you want to increase your egg production, will be your table birds, filling your own deep freeze for next winter's dinners or being sold to local hotels or butchers.

Those ducks you decide to keep for laying or for future breeding can be equally profitable, for a good, healthy bird will continue laying for as long as seven or eight years.

Yes, for the backyard farmer the duck is an attractive proposition.

4 Turkeys

Although I have bred and reared ducks and hens for more than forty years my favourite bird for profit is still the turkey. Like the duck it is a fast grower, it is relatively easy to keep and its flesh is still the most highly prized of all table poultry.

It came originally from Mexico, and it is said that when Cortés landed there early in the sixteenth century he found millions of them walking about, roosting and making themselves at home in the royal palace, and was able to buy prime birds for only four beads each. They would cost rather more now!

The turkey still runs wild in many parts of America, and the American Mammoth Bronze, which is first cousin to these wild birds, is one of the most popular breeds in this country. Stags, or turkey cocks, will weigh up to 40 lbs (18 kg) – wild stags may weigh up to 60 lbs (27 kg) and the hens about 18 to 25 lbs (8 to 11 kg).

The White, or White Holland, weighs slightly less but is equally popular, while the Norfolk Black is lighter still – only 10 to 16 lbs (4·5 to 7 kg) at maturity – but is particularly hardy and a good breeder.

HOUSING

Housing for your turkeys will be similar to that provided for the chickens, only larger. Turkeys should be kept in a large fold or enclosure, with a roosting house for shelter at night and with nesting boxes filled with straw. Don't forget that turkeys in the natural state roost in trees, so they will need perches and dropping boards, in fact the house will be very similar to the hen house we described in chapter 2.

BREEDING

If, as I hope, you decide to keep turkeys you can either buy day-old turkey chicks, or poults as they are more usually called, or month-old growers, or you may decide to breed them. Most poultry farmers would recommend the day-old poults as being far less trouble, but if you want to try your hand at hatching, with the help of either a broody fowl or an incubator, you will have to buy fertile eggs from a reliable dealer. (Unfortunately the mating of turkey stags with hens, always a troublesome business, is out of fashion.)

Having introduced the eggs to a broody fowl – about ten is all she can cover – or to an incubator set at 39·4° C (103° F), the procedure is much as for the hatching of hens' eggs except that the incubation period is longer – twenty-eight days instead of twenty-one. Use, if possible, a broody hen which has sat before, as a young hen may be put off by those extra seven days.

Turn the eggs yourself twice a day by hand while the hen is off the nest and, a precaution which applies to all hatchings, make sure that she is not troubled by lice or other insects. Treat her with a proprietary dusting powder or a poultry spray, paying particular attention to the feathers under the wings and round the vent. A hen, driven to distraction by mites, will often break eggs or scatter them over the nest edge as she attempts to scratch.

TURKEY POULTS

When the poults hatch leave them until the following day to dry off and gather their strength before moving them to a coop and a covered run. A turkey poult hates cold wet weather and is very easily lost in the first few days unless you keep it warm and dry. The covered run will keep the rain out, the mother hen will provide the warmth. However, if you have used an incubator you will find that the young turkeys take less easily to the brooder, so it will pay you to watch them for

a few days and, if necessary, shepherd them into the brooder
when they need warmth.

FEEDING

Turkey poults need plenty of feeding, for they grow quickly,
but if you should feel dismayed by the amount and cost of all
the food you prepare remember that the more they eat the
quicker they will grow, and the butcher or supermarket
manager who buys your table birds will pay you by weight.

What you feed to the turkey poult is, to an extent, a matter
of preference, although there are certain necessary ingredients.
For myself, I swear by an early diet of chopped hard boiled
eggs, chopped lettuce and bread crumbs. In a few days I start
mixing in some turkey crumbs, which you can buy from any
corn chandler, and then boiled nettle heads and chopped onion
heads for a change. But if these foods are not readily available
you will still get excellent results by feeding turkey crumbs
containing special antibiotics, which will prevent and control
diseases likely to affect young poults.

The turkey, more than other chicks, needs a good deal of
attention during those first few days, but it is, or will be, a
valuable asset, so it pays to make sure that it is warm, dry and
well fed.

As the birds grow, a big fold, similar to the one I recom-
mended for the fowls, only larger, is ideal. For ease of movement it should be on wheels so that you can push the whole
contraption of turkey house and run from one plot to another
every day. To reduce the risk of disease it is wise to dust the
used plot with lime.

As the birds grow – and they will grow quickly – they will
need even more food. Given a free run of your plot they will
pick up a lot for themselves, such as seeding grasses, docks,
dandelions, leaves and even the seed of thistles. Later in the
year they will find beech nuts and acorns and, if you have an
orchard, windfalls from the apple trees.

Confined to a run, a lot of this fresh food must be supplied
by you, although the birds will still forage among the leaves
and grass. Basically the daily diet for each three-month-old

turkey should be approximately 5 ozs (125 g) of grain and meal and 14 ozs (400 g) of green food: at six months these amounts should be doubled. In season, chopped turnips, potatoes and kitchen waste, cooked and dried off, with some laying mash will make a cheap feed, while if you have any skimmed milk from the house cow which, in a later chapter, I hope to persuade you to buy, this will make another useful ingredient. At other times a good home made diet might consist of boiled kitchen waste, plate scrapings, bread, crusts and flaked maize or turkey meal. It sounds complicated, but you will soon get used to it, and, even more, you will realize that, provided the basic ingredients are there, your turkey (or duck or hen for that matter) will thrive on what would otherwise have been wasted.

TURKEY BLACKHEAD

If you have been talking with neighbouring farmers you will almost certainly have heard of turkey blackhead. This was once a killer disease of turkeys but today is safely controlled by drugs.

Commercial manufacturers of animal foods will include the correct amount of antibiotic in the feed and thus alleviate the danger of this and other disease.

Well, I think that is all I can usefully say about my friend the turkey. So much knowledge, of course, comes from experience, from your day to day contact with livestock. But turkeys, you will find, are profitable to keep and do not need all that much attention. Their flesh is always in demand and not, as you might suppose, only at Michaelmas and Christmas. They would be well worth a try.

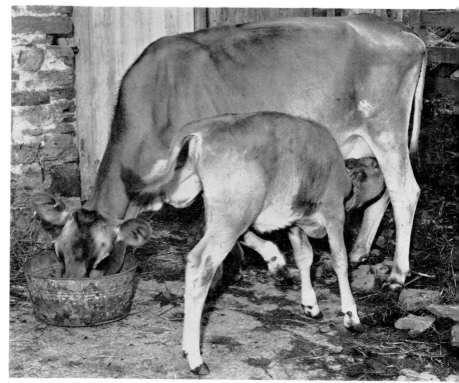

Above. My delightfully docile Jersey cow, Buttercup, suckling her calf.

Below. Lambs in springtime – in this case Ryeland lambs – are always entrancing.

Above. A rearing pen for poultry.

Below. Perhaps the least troublesome of all domestic poultry – the Guinea Fowl.

Among the most popular breeds of domestic fowl are the Rhode Island
Red (cockerel, *above*) and the Light Sussex (*below*).

Although in Australia the Muscovy Duck is a popular table bird, in Britain it is kept mainly for show – but any breed of poultry would thrive in this pleasant orchard setting.

5 Guinea fowl

Perhaps the least troublesome of all domestic poultry is the guinea fowl or galeeny. Originating in West Africa where it is still to be found in vast numbers in the wild state, it was known to the Greeks and Romans, disappeared for a century or two and was reintroduced to Europe some time in the sixteenth century when it was referred to as the turkey hen. It is a close cousin of the partridge and the pheasant.

Three domesticated varieties are kept in this country, the pearl, with purplish grey plumage and white markings, the white, and the lavender, which has a light grey plumage marked with white. The pearl, sometimes called the grey, is the most popular.

For your backyard or small-holding I can thoroughly recommend this handsome and courageous little bird, both for its market value as a table bird, for its eggs – perhaps a hundred or so in a year – and for its remarkable qualities as a watch-dog.

Yes, this little $3\frac{1}{2}$ lbs (1·6 kg) specimen of the family *Numididae*, with no spurs and, in fact, armed only with its beak and a courageous temperament, is one of the best early warning systems you can have. With Christmas approaching you need have no fear of prowling poultry thieves or tramps, for let them set so much as a foot inside your yard and your feathered pal will screech to make the heavens fall, 'Come back! Come back! Come back!' You will have no need for a dog.

A friend of mine had an excellent watch-dog, a border collie with ears so sharp that they could pick up the crunch of feet on gravel where the farm track left the main road half a mile away. Many a time have I walked up that track to the accompaniment of barking which grew louder and more urgent with every step. No one could approach that farm, I thought, in secret.

Well, I was away for a month or two, looking after my zoo and various pets' corners and making, at that time, perhaps two or three television appearances a week, and when I came to call on my friend again I noticed a change.

As I left the road and started up the track I walked in silence. Soon, I thought, old Gomar will bark, but no, all I heard was the crunch of shoes on the gravel, the wind in the branches and the skittering of small creatures through the hedgerow.

Gomar's dead, I thought. He was an old dog although not so old. Perhaps he had sickened and died while I was away.

Then, when I was only a quarter of a mile from the farm the silence was broken as first one, then another, then a whole chorus of screechings filled the air. 'Come back! Come back! Come back!'

My friend had the door open to meet me as I opened the gate, although it was quite dark.

'What's happened to Gomar?' I asked. 'Is he dead?'

'Not dead,' my friend said, with a chuckle, 'but retired.'

'Retired?'

'Aye.' He pointed to the barn where in the half light I could just see the old dog stretched out, head on paws, ignoring, or pretending to ignore, the dozen or more guinea fowl that were still protesting about my presence. 'Best watch-dog I ever had,' my friend said, 'until they came. Poor Gomar! He tried to compete with them at first but soon gave it up as a bad job. I reckon he's taken heed of the saying: "No point in keeping a watch-dog and barking yourself." '

You will find guinea fowl on many farms in the country, partly at least because they are so cheap and easy to keep. Given free range they would need virtually no extra food, although they will usually join in when the hens are fed.

If your soil is poor or if you have to keep your guinea fowl in runs, even when they have passed the growing stage, they will, of course, need feeding but it will not cost much, for you will find that they thrive on a diet of boiled household scraps, bread, potatoes, outside cabbage leaves and so on, mixed with mash. And don't forget the grit and shell which all domestic birds need.

But on most farms and, I suggest, on your small-holding, it is far better to let them run free. You will have to clip one wing of each bird before you let them out for the first time and repeat the process when they moult. This is a perfectly simple process, as natural and painless as cutting your hair, but it is done to throw the bird off-balance if it decides to fly away. Since they are adventurous little creatures it is a necessary precaution.

On free range, even on a well grassed half-acre (0·2-hectare) plot, they will pick up seeds, insects, weeds and so on, and will need very little extra feeding. If they join in, as they most certainly will, when the hens or ducks or turkeys are fed don't shoo them away for they eat little, and the bigger birds will be quite able to look after themselves.

In short, a dozen or so guinea fowl given free range of your poultry plot will provide a profitable and trouble-free side line. They are hardy enough to stand most cold weather, they need little feeding, they are immune to most poultry diseases although it will pay you to spray them from time to time against mites and other insects.

The hens, which are hard to distinguish from the cocks except for their characteristic cry 'Come back! Come back!', will lay about a hundred eggs a year, starting in March, and although you will have to search among the undergrowth and hedgerows to find them they make good eating and can form a valuable addition to your larder.

As table birds, too, they are well liked by hotels and supermarkets, and although the greatest demand is in February and March, outside the game season and before the spring chickens appear, you can in fact sell them all the year round.

BREEDING

Having kept these likeable little birds for a time and appreciating their qualities you may wish to breed them.

Well, this isn't difficult, especially if by this time you have had sittings of hens' eggs, for the procedure is much the same.

It is advisable to have a breeding pen with a cock and four or five hens, and when the fertile eggs are laid to transfer them to a foster mother. This is necessary because the guinea hen, although a friendly little bird, does not take kindly to man-made nests and unless she can make her own nest in the hedges will probably lose interest.

Once again a broody domestic hen is the answer unless you want to try your luck with an incubator. A broody hen can cover sixteen to eighteen eggs quite easily and once she has settled on her clutch you will only need to follow the procedure outlined in previous chapters. Hatching should be complete on the twenty-eighth day.

As an interesting side note you may like to know that many gamekeepers favour a cross Malayan Silkie bantam as a foster mother, for although she can only cover a dozen guinea fowl eggs she is a really excellent mother. If, at a later stage, you decide to add bantams to your farmyard stock, you might remember this.

REARING

When the guinea fowl chicks have dried off and begin to take an interest in their surroundings you may find that you have more of a handful than you bargained for. Domestic fowl chicks are lively enough, guinea fowl chicks are perpetual motion. They never stop, not for a minute. And adventurous! Any small gap in the netting or declivity in the ground and they are away.

So, being forewarned, you should have your coop and run ready: use ½-in (1 cm) wire netting, not only on top and sides but on the floor of the run, for it won't take them long to discover a way out. If you shift the run a few yards daily they will have fresh earth and grass to peck, and you can almost watch them grow.

For an early diet I recommend, as I did for the turkeys, a mash of chopped eggs and bread crumbs with lettuce leaves or onion tops, supplemented in a few days by pheasant crumbs.

At about a month or six weeks when they are ready to leave the run and join their big brothers and sisters outside remember to clip their wings and to spray them, while you still have them under control, against mites and parasites.

You may well decide after you have kept guinea fowl for a season or two that they are so easy to look after, so free from disease, and, let's be honest, so profitable, that you will double or even treble your present stock. Well, you couldn't share your plot with more agreeable neighbours.

6　Geese

Having used up most of my superlatives on the guinea fowl what can I say about the goose? It is hardy, easy to look after, singularly free from disease and, provided you have a paddock or a grassy plot, costs almost nothing to feed. It is intelligent, a conscientious mother, and is as good, if not a better, watch-dog than the guinea fowl. It is also the first recorded domestic bird.

Geese are mentioned in Sanskrit writings; it was geese which gave the alarm that saved Rome from the Gauls: the Egyptians had domestic geese more than 4000 years ago. In my part of the country and indeed all over Britain geese have been kept as domestic birds for hundreds of years, viz Tavistock and Nottingham Goose Fairs, and 'one fat goose' was a familiar tithe paid by peasants in the Middle Ages.

There must be a good reason for this popularity and, as I hope to show in this chapter, it is not difficult to find.

The most usual breeds of geese found in this country are Toulouse, a large grey bird, Embden, which is white, the Chinese or Swan goose, which is smaller but a more prolific egg-layer (it is also the most vociferous watch-dog of the lot), and the Roman and Sebastopol, which are smaller and white.

Perhaps I should also mention that in the west of England where I come from, and also in Canada, they have a sex-linked breed, that is, the goslings can easily be sexed at birth, but I feel that for the backyard farmer this is of academic interest only.

In fact I would say quite confidently that the breed for you is either the Toulouse or the Embden. They are both big birds and quick growers and therefore the most popular of all the table geese.

Now the first query which will probably cross your mind is whether geese need water – water to swim in, that is: all living creatures need it to drink. Well, my answer is the same as I gave for ducks. Geese don't need to swim, in fact they can exist quite happily and healthily on dry land, but if you intend to breed you will do so more successfully if the gander and geese have access to a stream or a pond.

So, let us assume that you decide to try your hand. There are several courses open to you, as for any other domestic bird, but you can probably do no better than purchase month-old goslings. Five, I suggest, four geese and a gander. They will settle in quite easily, you will find, and the only shelter you need provide is a simple shed.

So long as your poultry plot or paddock has plenty of grass they will fend for themselves, at least during the summer months, for they are grazing birds, and you need not feed them further except for the occasional treat of cabbage leaves and handfuls of corn. In an exceptionally dry summer or in the winter when the grass has been well cropped you should supplement their diet with cabbages or other greenstuff, boiled turnips, potatoes and swedes, stale bread and kitchen waste and a few handfuls of grain. Also, like other laying birds, they need a supply of grit and shell.

You will soon have to decide whether you want to keep your small flock for breeding purposes or whether you prefer, or perhaps need, to take the quicker return by rearing for table meat.

Most farmers rear two batches of geese a year, one hatched in early spring for the Michaelmas trade, one hatched in midsummer for Christmas. In either case the profit margin is excellent, as you will soon see, for if your half-acre (0·2-hectare) poultry plot has plenty of grass and a water supply they can be virtually self-supporting. Your only problem will be how much extra food to give them and how far this will be reflected in extra poundage.

BREEDING

But if things are going well you may decide that it will be more profitable in the long run to breed your own goslings, thus deferring for a few months at least any 'table meat' profit, or you may decide to sell one goose for table meat and keep the rest for breeding.

Well, unlike the turkey, which can be something of a caveman lover and has no interest in its offspring, the gander is the perfect gentleman. He won't mate until he has got to know his 'intended' quite well – usually five or six weeks – and he will protect the nest and the goslings, when they are born, with courage and ferocity.

So, if you are intending to breed you should allow at least five to six weeks 'engagement' period, during which gander and geese can feed together and become acquainted. In the meantime you should provide ample bedding, straw for preference, so that the geese can make their own nests in quiet corners of the shed.

Fortunately, the goose, like the gander, makes an admirable parent, so, for once, you won't have to go looking for a broody hen (in fact the hen is seldom happy with a clutch of goose eggs since they are too big for comfort, and the incubation period of thirty to thirty-one days is much longer than she would naturally expect).

When the goose has chosen her nesting spot, make sure that she is sheltered from rain and unnecessary draughts and that the eggs are protected by wire netting against foxes and rats. In fact if you feed her once a day where she can keep an eye on the nest she will soon send packing any marauding birds such as crows and magpies and will even give short shrift to a rat. If a fox appears she will scream for the gander and, as I have seen for myself, it is a very hungry fox which will stand up to such formidable adversaries.

The goose will cover up to fifteen eggs, but ten or a dozen is more usual: any spare eggs will find a ready market.

The eggs will hatch in a month (thirty to thirty-one days) and if you are feeling impatient and perhaps wondering whether the mother goose is going to bring it off you will be reassured towards the end of the period by the faint but unmistakable sound from inside the eggs, for goslings, which really are the most delightful of creatures, cheep away while they are hatching and thus come cheerfully into the world.

Goslings hatched naturally are generally stronger and less

troubled by illness than those hatched by incubator. With the gander to protect them and the goose to teach them to eat and swim and fend for themselves they make rapid progress. Feed them on crumbly, not wet, chicken mash, and in a few days you can add some finely chopped lettuce leaves. It won't be long before, following mum's example, they are nibbling at fresh grass.

THE INCUBATOR

If, in your second or third year, your flock has increased so much that you can't find space or time to hatch all your eggs naturally you can get quite good results with an incubator. Suppose you have forty to fifty fertile eggs or even more; then incubator hatching can be a commercial proposition. Don't worry about what you are going to do with so many goslings, even if you only have a half-acre (0·2-hectare) plot: fourteen-day or month-old goslings can be easily sold and, since they have cost very little to feed, this is a most profitable operation.

When you decide to use an incubator I suggest that you pencil the date on each egg as you collect it: in this way you will ensure that only the freshest eggs are used. The incubator should be set to a temperature of 38·5° C (101° F), which can be increased to 39° C (102° F) towards the end of the thirty days.

Turn the eggs daily, not just sideways but end to end, and cool them for ten minutes a day by leaving the door open for this period from the twelfth to the sixteenth day, and for sixteen minutes a day from the sixteenth to the twenty-sixth day. In very dry weather it is helpful to spread a piece of felt soaked in warm water over the eggs for a few hours to assist hatching.

Inspect the eggs regularly – the best time is when you turn them – from the end of the first week. Fertile eggs will show a dark blob. Remove any infertile eggs as they will give off fumes injurious to the others.

When the goslings hatch they will need a fair amount of attention until they find their feet. Give them a run in a fenced off part of a draught-proof shed, with an exit to a small feeding area where they can pick up the chicken mash and finely chopped lettuce leaves and, before long, young blades of grass. In fact it is a good idea to give them a few fresh turves dug from the paddock and to replace them every few days as they get dry.

Don't assume that because they are geese they need plenty of water. In fact at this age, without a mother to look after them, they catch cold very easily, especially if they fall or venture into too deep drinking bowls or get soaked by rain or long wet grass. So give them a shallow drinking bowl with a largish stone in the middle to restrict the area into which they might fall. It is also a good idea to lay out an old hessian sack in their feeding area for the first few days as this gives them confidence and keeps them dry.

And clean, of course: all poultry must be kept clean and dry. Put a layer of granule peat in their run and change it frequently. It is these small points, which are so important but which can so easily be overlooked.

Remember, too, that these goslings represent part of your assets: whether you sell them in a few weeks as growers or in a few months as table birds or whether you decide to keep them to augment your breeding stock they will bring you profit; what is more, they will prove, if you still need convincing, that backyard farming can be a viable proposition.

7 Pigs

So far in this book we have dealt with poultry, chickens, ducks and guinea fowl and other birds which are easily manageable by anyone with his heart in the farming experiment.

Now we are coming to the heavy brigade, heavy and expensive perhaps for the beginner, but I hope that with a measure of success behind you your nerve won't fail.

Pigs and sheep and cows are the very stuff of farming, and in case you feel that they can only be kept by the expert just think of the cottagers and small-holders you have seen, or perhaps known, with a pig or a sheep or perhaps a milking cow in their backyard. It only needs courage and hard work.

So, with those reassuring words, let me introduce you straight away to the pig.

It used to be the most profitable of all animals for the small farmer. Now, owing to the cost of feed stuffs, the EEC green pound and a host of other factors, this is not so. Yet the pig is still the pig, prolific, quick growing and always in demand. Oestrum, or coming on heat, occurs every three weeks with the sow, except when she is suckling; the gestation period from mating to birth of the litter, called farrowing, is sixteen weeks, and since oestrum recurs within a fortnight of the piglets being weaned (allow eight weeks for weaning) the whole reproduction cycle can be completed in six months. It is therefore possible and indeed usual for the sow to produce two litters a year.

How many is that? Well, it depends on the breed – some sows are notably more prolific than others – and to an extent on the age and health of the sow, but if you reckon on an average of nine piglets to the litter, of which eight will survive, you should get about sixteen piglets a year per sow. (Some sows in fact will produce almost twice as many, but let's keep our expectations modest.)

HOUSING AND EQUIPMENT

The next thing you will want to consider is housing and equipment, and to do this it is necessary to understand certain characteristics of the pig.

In the first place it is not, as is generally supposed, a dirty creature, in fact quite the contrary. Given enough room, it will confine its droppings to an outside run or to one corner of the sty, keeping its sleeping quarters clean. Too often you see sows wallowing in their own filth: when you see this blame the pigman, not the animal.

The pig is also susceptible to draught and damp and will suffer more than other farmyard animals from extremes of cold and heat. So whatever housing you decide on make sure that it is strongly made, warm and dry in winter and not too hot in summer.

The pig is also very strong, particularly when it uses its considerable weight, so don't expect to contain it in any rickety old shed, at least not for long.

Now, depending very much on the state of your finances you must decide whether you will have permanent or more temporary housing.

Permanent pig house

A permanent pig house of concrete or breeze blocks faced with cement is ideal, although quite expensive unless you build it yourself. Erect it on level ground, facing south to catch the sun and make sure that there is adequate drainage. Be sure, too, that you have easy access for there is nothing more annoying than having to adopt a Yoga position, standing on one leg with your head level with your knees as you attempt to use a muck rake with one hand and fend off a 4 cwt (200 kg) sow with the other. Incidentally, the pig house should be cleaned out daily.

If you are going in for breeding, then your permanent

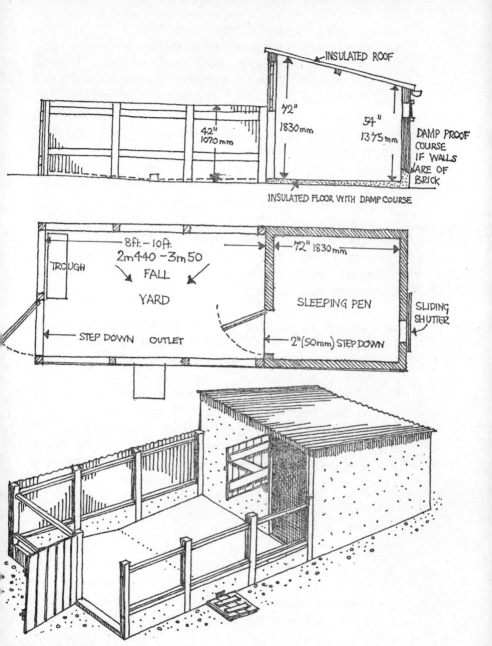

INSULATED ROOF

72"
1830mm

54"
1375mm

42"
1070mm

DAMP PROOF
COURSE
IF WALLS
ARE OF
BRICK

INSULATED FLOOR WITH DAMP COURSE

8ft.–10ft.
2m440–3m50
FALL

YARD

TROUGH

72" 1830mm

SLEEPING PEN

SLIDING
SHUTTER

STEP DOWN   OUTLET

2"(50mm) STEP DOWN

sign for a two-pig sty

house needs to be rather more elaborate. A typical house would consist of a sleeping compartment for the farrowing sow and her litter and a run. Both would be roofed over and completely enclosed during inclement weather, although you need vents under the roof for an adequate supply of air.

Now, as we consider the inside of the sleeping compartment, there is another characteristic of the pig, or rather the sow, which we must remember.

They are, of course, hardly lightweights, however quickly they may twist and turn when you try to catch them, and, with such short legs, there can be no question of lowering themselves gently to the ground. When the sow is tired of standing she will simply subside and roll over.

Unfortunately she doesn't always look where she is sub-siding and unless you are careful you may lose half your litter – 'squashed by mum'.

To avoid this you either need a creep, which is a fenced-off area where the piglets can rest in safety, or a fencing rail. This latter is erected about 3 ft (1 m) from the wall, and the area so guarded where the piglets can lie is usually heated by an infra-red lamp. Many piglets die from cold, for they have little natural covering to keep them warm, so by this method you protect them from both exposure and squashing.

In fact after your sow has had her litter you will appreciate the homely scene: mother lies in the middle of the pen where she can do no harm while the piglets snuggle under the lamp only 'coming to life' when they wish to suckle.

Temporary houses

Temporary houses can be of varying shapes and sizes and will depend very much on the material you have, the size of the plot and on your own ingenuity. Having said this, anything you construct should conform to the principles set out above, that is, it should be warm and dry in winter and not too hot in summer.

You may have an existing shed which you feel might be adapted, and this could well be satisfactory provided it is warm and rain-proof and of sound construction. Don't waste

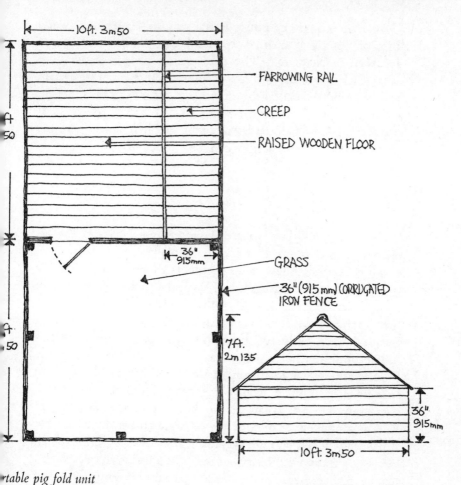

table pig fold unit

your time with a rickety shed which has seen better days, for the pigs will have it to bits in no time.

In my part of the country we favour a pig ark, a simple construction, about 6 ft (2 m) high at the apex, and with a floor space approximately 8 ft by 10 ft (2·5 m by 3 m). The sow usually wears a harness restricting her movement outside the ark with the tether long enough to give her free range over an area of grass but not so long that she ties herself up into knots – as she will, given half a chance. The ark, which should be made of heavy timber, is on runners so that it can be moved from plot to plot.

Another temporary house is the pig fold, which is on the lines of, although more substantial than, the hen fold we described in chapter 1. The fold unit, which is particularly suitable for a sow and litter, is a house about 10 ft (3 m) square with a wooden floor and a creep or farrowing rail. Outside, a run of approximately the same area, that is 10 ft by 10 ft (3 m by 3 m), is enclosed by a corrugated iron fence. Like the ark the whole construction can be moved to fresh pastures when required.

Feeding troughs

The pig, as you know, is a greedy and messy eater and it is not always easy to ensure that it gets the full benefit of the food you give it or that the weaker members get any at all. Movable troughs are seldom satisfactory, since the pigs will have them over in no time and much of the food will be trodden into the ground in the resulting scramble. (In fact the pig doesn't miss spilt food easily. With its powerful snout it will root and snuffle until any last crumbs have been recovered, while the food splashings on a wall or on a door will probably result in splashings and door or wall being gnawed together.)

If you have a permanent pig house you will do well to have a trough, either of concrete or stout timber, built into the wall. Make it big enough to ensure that there is room for all the pigs, otherwise the smaller and less aggressive ones won't stand a chance. Arrange it, too, so that you can pour food into the trough from outside. I have been knocked flying more than once by ravenous pigs and have considered myself lucky if I didn't land in the mire with a generous helping of pig swill in my lap.

My own preference is for old kitchen sinks, which you can often buy quite cheaply from builders' yards, and which are generally too heavy for even the most hungry pig to upset. Don't forget to put a wooden plug in the hole at the bottom of the sink.

For your farrowing pen where you only have the sow to consider I recommend a galvanized iron bucket which makes it easy to measure out the correct amount of food and which will not result in much, if any, food being lost if the sow knocks it over.

Other equipment

Something you will have to consider is how you are going to handle a 3 or 4 cwt (150 or 200 kg) pig once you have reared it. In the first place how will you know how much it weighs?

Well, if you can afford it, you should buy some kind of weighing machine. You won't be able to extend to the weighbridge such as they use on all the big pig farms, but there are smaller machines, rather like the parcel weighing machine used by British Rail, which you can sometimes buy second-hand at agricultural dealers or at markets. I am not saying that a weighing machine is essential, but it is obviously an advantage to know the weight of your litter and of any mature pigs you want to sell.

And talking of selling don't forget that you will need to keep a record of all pigs bought and sold. This is the law!

Now, how are you going to get your pork or bacon pigs to market? They will be too heavy to lift into the van, even if there were anything to catch hold of. And how are you going to get your bigger pigs on to the scales?

Well, the answer is a ramp, a mound of earth fenced on two sides, up which your protesting livestock can be driven. I use two sheets of galvanized iron to guide them into the van. If you have two levels, one the height of your weighing machine, the other to the floor level of your van, you will save yourself a lot of frustration and bad temper. I know!

One other item which we haven't considered yet is fencing. The ark, we said, needed no fence because the sow was in harness, while the run of a fold unit is enclosed by stout timber and corrugated iron.

But supposing you have bigger runs, which will be essential for your rearing stock. How are you going to contain half a dozen or more pigs?

Well, the best, but the most costly, way is with fences of pig wire, a strong, large-mesh wire netting, which will withstand the considerable weight of mature sows and pigs. As an

C

alternative you can use a single or double strand electric fence.

Home breeds

Now, before we get down to discussing the whole cycle of breeding, weaning and fattening for pork or bacon it will be helpful to consider some of the breeds available.

Pigs, of course, have been part of the British scene for as long as history is recorded. The wild pig, from which our present breeds are descended, was still roaming the fens of East Anglia as late as the sixteenth century, but although the swineherd tending his herd in the forest was a familiar figure in medieval times it was not until the second part of the eighteenth century that the domestic pig really came into its own.

There are now thirteen recognized breeds in Britain, but since many of these are generally confined to certain areas we need only concern ourselves with a few.

The most popular breed of all is the Large White, which, with its long body and well developed hams, is an excellent bacon producer. It is a quick grower, and, from experience, there is no other breed I would sooner recommend. It is one of the older breeds, having developed from the earlier Large White Yorkshire, and there is a record of a prize beast late in the eighteenth century weighing over 12 cwt (600 kg). (Compare this with the heaviest breeds today which seldom exceed 5 cwt [250 kg].)

Because it is such an excellent bacon pig the Large White boar is used extensively for breeding and is frequently crossed with sows of other breeds to produce a litter of white, quick fattening baconers. The most usual cross is with Landrace, the most important of the Continental breeds, and a large proportion of modern pigs are hybrids of this combination.

The Welsh breed, another bacon pig, is also extremely popular and with the Large White/Landrace cross accounts for a major portion of pig stock in this country today.

British Saddlebacks are good dual purpose pigs although generally considered inferior to the Large White for bacon.

The sows make excellent mothers and rear large litters, so this is another breed you might consider. Crossing a Large White boar with a Saddleback sow produces good sturdy litters which are much favoured for bacon production.

PORK OR BACON

Now, having considered the main breeds you will have to decide whether you want to go in for pork or bacon. Some of the above breeds are dual purpose, as I explained, so if you buy a Saddleback your decision can be deferred, but I think you should understand from the start the difference between porkers and baconers.

To be successful as a breeder you must have your pigs ready for killing, for whichever market you choose, in the shortest time. A porker should be ready at fourteen to sixteen weeks, a baconer at twenty-two to twenty-six weeks.

The demand for bacon is more or less steady throughout the year, so your baconers can be killed at any time. Pork is not much in demand in the summer, so you will have to avoid, if you can, having to kill porkers in that period.

BREEDING

Now I hope you are not getting impatient, but there is one other important aspect of pig farming you should consider before you go out to buy your first sow. You must understand something about breeding.

First, there is the gilt, as she is called until she is mated and produces young. Later she will be called a sow. She should be sound and healthy and bear all the characteristics of her breed. She should have at least twelve and preferably fourteen sound teats, or drills as the pig breeder calls them. She should be docile and even tempered. She should be at least $7\frac{1}{2}$ months old and weigh around 260 lbs (120 kg) when she is first mated, and will be nearly 1 year old when she farrows.

The boar, of course, is equally, or perhaps even more important, but as I am going to suggest that on your backyard farm you will be well advised to start with a sow in-pig (pregnant) we need not consider the boar at this stage. Later, when your sow has had her litter, you can re-mate her with a neighbour's boar, or use artificial insemination (AI).

We will consider later the amount of food a sow requires before and after farrowing, but it is worth mentioning here that the pregnant sow should be allowed plenty of exercise. Put her out to pasture in the summer, but remember that her sleeping quarters – and this applies to all pigs at whatever stage of their development – must be warm and dry. Pigs kept in damp or draughty quarters suffer from rheumatism and easily catch chills which can be fatal. In winter you will be well advised to confine your pregnant sow to the yard.

Three days or a week before farrowing the sow should be moved to her farrowing pen.

The size of the litter may be anything from four to fourteen, but you should be well satisfied if your sow produces eight or nine. It is best if only yourself is present during the farrowing as it is essential to keep the sow quiet. In fact the first couple of days after farrowing is the danger period when the mortality rate among piglets is highest, and a nervous sow can easily lie on several piglets, and crush them to death.

If your sow proves to be extra prolific remember that she can only cope satisfactorily with as many piglets as she has teats, and in this case the surplus can be transferred to another sow who has been less prolific, although in the early stages of your farming experiment you may only have the one sow.

In some litters there may be a runt – a piglet smaller and less active than the rest. Unfortunately the runt will be bullied by his bigger and stronger brothers and sisters and will always be small and from a commercial point of view quite unprofitable. You may like to hand rear it. This is pretty simple – treat it like rearing a small puppy, feeding it with some powdered milk and baby food to start with.

At Little Comfort more than one piglet was brought up in the kitchen, fed with a bottle and treated as a family pet. Delightful little creatures they were, pink and cuddly and, when they were first brought indoors, rather pathetic. Unfortunately they never became house trained!!

Provided your litter survives the first couple of days and has warm dry sleeping quarters and enough to eat, it will grow quickly and at the end of eight weeks should be weaned. Don't be tempted to leave the piglets with their mother any longer as this would upset the whole breeding, rearing, selling cycle and make considerable inroads into your profit.

The ratio of male to female piglets varies, and you will of course hope for fifty per cent or more being females. The males, unless you intend to keep one for breeding, should be castrated a few weeks after birth; they will be termed hogs.

FEEDING

Sufficient correctly proportioned food is more important for pigs than for any other farmyard animal, because unless you feed pigs correctly the profit margin, which will be highly satisfactory if you do the right things at the right time, can be quickly whittled away. A gilt or a hog which has to be kept and fed for longer than the normal period, that is fourteen to sixteen weeks for a porker or twenty-two to twenty-six weeks for a baconer, will cost you money. Pigs in prime condition and ready for killing at the right time must be your object.

So, how should you go about it?

Let us consider first the growing pigs after they have been weaned. Because they are greedy creatures and appear to relish anything which is tipped into their trough it may surprise you to know that feeding pigs is a highly complicated business. You can't feed them large quantities of cheap, bulky foods such as hay and greenstuff, which would satisfy many of your farmyard animals, but must give them a controlled diet which includes a proportion of protein foods such as fish or meat meal or other animal protein food. Complicated, I said: well, listening to professional pig farmers you may think it a bit too complicated, and you could be right. Some pig breeders I know keep elaborate wall charts showing the exact amounts of food, with proportions of protein to bulk according to the pig's age and weight. You may think that without resorting to haphazard feeding you could manage quite successfully with something more simple.

Basically the young pig needs about 10 per cent protein

food, reducing to 5 per cent when it is four months old. The protein food will be fish meal or meal from other animal sources, and the proportions can be reduced if the diet includes skimmed milk. (If you keep a milking cow this should present no problem.) Vegetable protein such as soya bean meal, is less nutritious but can be used as an alternative provided lime and salt are added.

To make up the bulk of their diet you can buy barley meal, weatings and maize (*see* Model Rations chart) but this, of course, is relatively costly and most backyard farmers cut their food bills by including with the meal such bulky foods as potatoes and swill. Potatoes or other root crops, which should be boiled or steamed, can be substituted in a ratio of four to one, that is, 4 lbs (2 kg) of potatoes to replace 1 lb (0·5 kg) of meal.

Swill is far less popular these days, partly because of the disease risk, partly because of the poor nutritional value, but if you are producing a pig for your own table it is a worthwhile substitute.

Swill, which may come from hotels or restaurants or army camps, *must* be boiled for one hour before serving, otherwise you will be breaking the law, and it may be worth making inquiries at your local council offices as some councils collect swill and process it for use by pig farmers.

Incidentally, if your pig food is mainly barley meal or a combination of barley meal and maize you will have one of the best fattening foods you can get, especially if some skimmed milk is added.

MODEL RATIONS

Ration for Suckling Sows and Growing Pigs	*Ration for Fattening Pigs* (over 120 lbs/54 kg live weight)
10% Fish meal	10% Soya bean meal
20% Weatings	10% Weatings
20% Flaked maize	20% Flaked maize
50% Barley meal	60% Barley meal

You may purchase proprietary rations called Sow and Weaner, Grower, Fattener and specially formulated 'Creep Feed' for suckling pigs.

To get some idea of the amount of food required you should give a pig weighing 40 lbs (18 kg), that is soon after weaning, about 2 lbs (1 kg) of meal a day, which will increase as he gets older and heavier to a maximum of about 6 lbs (2·7 kg), or roughly, feed 1 lb (0·5 kg) of meal per month of age (*see* Daily Allowance for Growing and Fattening Pigs).

DAILY ALLOWANCE FOR GROWING AND FATTENING PIGS

Age in Months	*Meal Daily lbs/kg*	
2	2/0·9	Grower meal
3	3/1·36	Grower meal
4	4/1·8	Grower meal
5	5/2·26	Fattener meal
6	6/2·7	Fattener meal

or

	Grower meal	*Cooked Potatoes lbs/kg*
2	2/0·9	—
3	3/1·36	—
4	3/1·36	4/1·8
5	3/1·36	8/3·6
6	3/1·36	12/5·4

After three months old I would suggest a basic 3 lbs (1·5 kg) of meal a day, plus high protein food, the balance being made up of potatoes or swill. You will soon learn, partly from experience but more probably from discussions with local farmers, what your pigs need to thrive. Don't be upset if you get varying kinds of advice: every farmer has his own pet theories, but I think when you analyse them you will find that they all agree on the basic principles set out above.

And, who knows? You might chance on some happy combination which beats them all.

Sows in pig

Now before we leave the subject of feeding we must not forget the sow in pig.

You will remember that I suggested that in summer she should be turned out to pasture, and this, in addition to giving her healthy exercise will cut down your food bill. Give her about 4 lbs (2 kg) of meal, including 10 per cent protein food, a day if she is grazing. If it is winter and she has to be kept in the yard increase this to 6 lbs (2·7 kg).

In the week before farrowing, when she is kept in the farrowing pen, the ration should include a proportion of greenstuff, or coarse bran to keep her bowels open.

After farrowing the sow will become even more hungry, especially if she has a big litter, and you may have to increase her daily ration to about 6 lbs (2·7 kg) for the sow plus ½ lb (0·3 kg) for each piglet (*see examples below*).

DAILY ALLOWANCE FOR SUCKLING SOW

Example: Sow and 12 piglets

Sow	6 lbs/2·7 kg
12 piglets × ½ lb/0·23 kg	6 lbs/2·7 kg
	12 lbs/5·4 kg daily

Example: Sow and 8 piglets

Sow	6 lbs/2·7 kg
8 piglets × ½ lb/0·23 kg	4 lbs/1·8 kg
	10 lbs/4·5 kg daily

Piglets

The young pigs will gain sufficient nourishment from their mother for the first two or three weeks, but they should be encouraged to eat solid food as soon as possible. To do this you will have to put a feeding trough in the creep (a 'pig creep' is a fenced off area where the suckling pigs can lie away from the sow and obtain supplementary food called 'creep feed'). From about two weeks onwards fill it with creep pellets or with a small quantity of meal. By eight weeks the young pigs will be weaned.

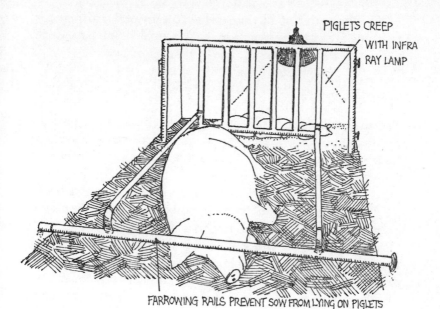

PIGLETS CREEP WITH INFRA RAY LAMP

FARROWING RAILS PREVENT SOW FROM LYING ON PIGLETS

Indoor farrowing pen

HOW TO START

So, having considered all, or most, of the problems you decide to try.

Well, as you have gathered from what has been said before, there are various ways of going about it, but for you, I suggest, there are only two. The backyard farmer, in my opinion, has to decide between establishing a small sow unit with all its attendant hazards such as getting the sow in pig, health and mortality risks before, during and after farrowing, and so on, and buying weaners for fattening.

The sow unit we have already discussed, and you will have gathered that the problems of farrowing and then raising the young piglets are considerable, especially for the beginner, although by no means impossible. If you decide to make this your plan you should think in terms of breeding to weaning stage, that is about eight weeks after farrowing, and then selling. Eight-week-old pigs find a ready market and, subject to the economic scene at the time, should find a reasonable price.

The alternative of buying eight- to ten-week-old pigs for fattening is much less hazardous and, provided you have done your costings right, should show a margin of profit.

Specimen costing

We shall talk a lot about costings, which are so important if you are going to make your farming experiment a success. Unfortunately, because of inflation and other economic uncertainties it is not possible to give actual costings since they would almost certainly be out of date by the time this book reaches the reader. As an example, however, it might be useful to explain how you should go about it.

Assume that you will keep a pig for twenty-four weeks, that is approximately from piglet to marketing stage. You have examples of typical pig fattening rations in Charts A and B. By making a few enquiries you can find the current prices and make a fair estimate of what food alone will cost. Cereal feed is expensive, protein, especially animal protein, more so.

You can then find the current average price of pigs, say £x a score, so if you sell a pig at 200 lbs (90 kg) you will get £10x.

The margin between £10x and the feeding costs, from which you should also make some allowance for housing, labour, light and the services of your vet, represents your profit. At the present time it is likely to be small.

A gloomy picture? Well, yes, it is, but as I indicated earlier, pigs are not one of the more profitable forms of livestock at the moment. You may think, and I would agree, that some of these feeding costs are on the high side. If you were fattening a pig for your own table you could certainly save a good deal by including in the diet boiled potatoes and swill.

Other costings may well be less gloomy, but they need to be made whatever livestock you are considering. The object of backyard farming, as we have said, is not large profit but self-sufficiency. Only an accurate costing will tell you which livestock to buy.

Slaughtering

One last word. Before you start fattening pigs for your table make sure that you have slaughtering arrangements laid on. The number of professional men who can expertly butcher a pig and dress it is getting smaller all the time. The High Street butcher is invariably linked to the big abattoir where animals are slaughtered on a 'line' basis. Some local inquiries are advisable.

DISEASES

Now for the dread subject of diseases. I say 'dread' because the pig is a relatively expensive animal and for any farmer a valuable asset, so it is necessary to pay close attention to health. Rheumatism we have already mentioned and chills which can be fatal arising from damp sties or pens.

After you have been working with animals for a time you will soon recognize the symptoms of disease. Always take heed if the pig is constipated or, even worse, if it scours, that is, suffers from violent purging diarrhoea. A sick animal will look listless and may be off its food, which, for a pig especially, is easily noticeable.

It may run a temperature, the normal being 38·9° C (102·3° F), constantly lie down and refuse to eat its food, have purple blotches on its belly, or have a discharge from the nose, eyes or anus. Should you see any of these signs of ill health, then call in a veterinary surgeon. Finally remember that a curly tail is a good indication of a healthy pig!

Foot and mouth disease

This is a highly dangerous specific infection caused by a filterable virus. Infected animals will develop a fever and may show lesions on the feet or snout, which will cause the pig considerable pain, making it difficult to stand. Pigs will be most distressed and found squealing.

Should you have the misfortune to suspect this disease you

should promptly inform your Ministry of Agriculture Veterinary Surgeon or the local police.

If the disease is confirmed then all your pigs, cattle and sheep will be destroyed by the Ministry and you will receive compensation.

Swine vesicular disease

This is very similar to foot and mouth disease but only affects pigs (*see above*). In this case if the disease is confirmed only the pigs are slaughtered.

Mange

This is caused by parasites in the skin and causes scabs and sores. If your pigs suffer from these look to your cleaning methods for they should not occur in regularly cleaned and disinfected sties. It can generally be cured by washing the infected pigs and treating them with a prepared powder.

Swine erysipelas – 'purples' or 'diamonds'

This disease is characterized by red or purple blotches on the skin – hence the name 'purples' or 'diamonds'. It is most likely to occur in the hot summer months. The pig goes off its food and runs a high temperature.

Prompt attention by a veterinary surgeon is necessary and usually the pig responds to treatment.

I hope the last paragraphs won't have put you off. Diseases are largely preventable by giving your animals the right foods, boiled where necessary, and keeping them and their sties and runs clean.

This has been a long and detailed chapter because – let's face it – pigs are not as easy to keep as, say, guinea fowl or hens, but there's nothing magic about it. Anyone with common sense and a bit of land, a good shed and a pair of wellington boots can make a success of pigs. Of course he mustn't be afraid of getting his hands dirty and he must be prepared to work.

Pigs are profitable, although not as profitable as they were, and a pig farm, properly run, shows a minimum of waste. I said earlier that a deep freeze was, if not essential, at least highly desirable, and you will be surprised how your family budget will benefit by the addition of a pig.

Unlike some animals the pig provides a carcase with scarcely any waste. You can make faggots from the liver, heart, lungs and head; black, or potten, pudding from groats mixed with blood; chitterlings, brawn made from pig's head and trotters and tails boiled and put in a gelatine mould. The countryman wastes nothing, nor should you.

Bacon can be cured with common salt in an earthenware tub called a salter. The process is quite simple but, unfortunately, is outside the scope of this book.

And talking about waste don't forget that pigs' droppings make the very best manure. Add them to your vegetable patch and you will achieve wonders.

And to finish, if I may, with one word of advice. Farming is an up and down business, by which I mean that you will have your good years and your bad. The wider your range of animals the more even will be your profits, on the principle of swings and roundabouts.

But if you sink a fair amount of capital in pigs do so with the idea that good years will follow the bad. There is a great temptation, which even experienced breeders find hard to resist, to follow the market and panic sell when prices are falling and buy when they seem likely to rise. As one who has farmed on and off for more than forty years I can assure you that things will even out in the long run. Go your steady way, with faith in your own ability backed by hard work, and you cannot fail.

8 Goats

Now from our serious survey of pig business let us turn to another aspect of backyard farming for light relief.

Yes, I mean our old friend the goat, and, anticipating protests from the fanciers, let me add that no one has more respect for the unsung heroine of the farmyard, the nanny, than myself. She is the hardiest of creatures and one of the most undemanding. Given free range, with perhaps a couple of others, on your half-acre (0·2-hectare) plot, with dry bracken for bedding in winter and supplementary feeding of hay, which you can cut from your own verges, and kitchen waste, she will provide you with milk and cheese, superior in my opinion, to any product of the cow, she will produce kids for your table or further breeding, she will clear all those thistles and brambles and unsightly weeds from your plot. And all this with a minimum of attention by yourself.

Goat keeping could hardly be more simple. All you need to know is that a kid becomes a buckling if he is male or a goatling if female at the age of one year and a buck or billy (male) or nanny (female) at the age of two. Apart from this everything is common sense based on the general principles which apply to all farmyard animals, such as warm, dry sleeping quarters, supplementary feeding in winter or in times of drought, cleanliness and a constant supply of drinking water.

MAIN BREEDS

The goat is a mountain animal which can do well on the most unrewarding pasturage (it considers gorse a delicacy) and can withstand a good deal of heat. It is found wild and in large domestic flocks in most Mediterranean countries, but the best milk producing breeds come from farther north.

Of these the most well-known are from Switzerland, the Alpine, Saanen and Toggenburg, although the Nubian has also contributed greatly to British stock. Thus the most successful breeds in this country are known as British Alpine, British Saanen, British Toggenburg and Anglo-Nubian, the original breed of Old English having been virtually eliminated by cross breeding.

As a start I suggest you buy a two-year-old nanny, which will need the minimum of attention and will prove, I hope, that goats really are worth keeping.

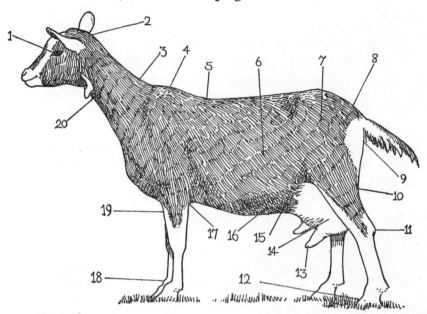

Points of a goat

1 Eye, bright and gentle
2 Head, shapely and intelligent
3 Neck, long, not coarse
4 Shoulders, clean and neat
5 Back, the line long and level
6 Ribs, deep and well sprung
7 Pelvis, wide
8 Rump, sloping gently
9 Escutcheon, wide and reaching high
10 Rear of udder, well developed
11 Hocks, wide apart and straight
12 Feet, sound and neat

13 Teats, pointed and directed forward
14 Udder, spherical and firmly attached
15 Barrel, ample for food
16 Milk veins, prominent
17 Body, deep allowing room for heart
18 Pasterns, fairly straight
19 Forelegs, straight, sound, not too close
20 Throat, clean and fine

HOUSING

First you will need somewhere to house her at night or in inclement weather, and in this connection I should mention that goats can more easily withstand heat than cold or wet. Nothing elaborate is necessary, simply a shed that is dry and free of draughts although still with an adequate air supply.

FEEDING

She will also need drinking water and, as with any other animal, this should be regularly cleaned.

The old-fashioned method, which is still used by many goat-keepers, is to provide an open drinker or earthenware trough. Others prefer a bucket held in position by the rim of a larger bucket nailed in a corner of the shed. Either method is satisfactory if, and only if, you keep the drinking receptacles clean. The bucket should be scoured and re-filled daily, the trough should be emptied, scoured and re-filled. The temptation, particularly with the heavy trough, is to slosh more water in when the level is getting low, not troubling about the dirt and parasitic germs which may be lurking in the mixture. Cleanliness, as you will probably hear me say many times in this book, is just practical common sense.

If you want a slightly more sophisticated contraption, one which I can thoroughly recommend, you should invest in a goat drinker. This is a galvanized iron tank or trough controlled by a ball trap so that the water cannot overflow. As the goat presses the tongue of the valve she releases an inflow of clean uncontaminated water, and all you need to do is make sure that the cistern is full.

Fix the trough about 12 to 14 ins (30 to 35 cms) from the ground so that she cannot foul it, and if at first she seems slightly puzzled wait until she empties the trough and then show, by gently pressing the valve, how she can help herself. She will soon learn.

Feeding, as you will have gathered, is no great problem, for the goat is not a fussy eater. If you watch your nanny goat you will see that she is a browser rather than a grazer and that she can eat and thrive on much coarser food than a cow. She is a ruminant and I think you will find, as I do, that there is something essentially peaceful and satisfying in the sight of a nanny tethered in the paddock or perhaps sheltering in her shed and chewing the cud. She will eat all sorts of rough vegetation such as weeds, thistles, brambles and gorse, but be sure to keep her from certain poisonous leaves and berries such as yew, and from young trees, which she will quickly strip bare. Also, if you want to preserve your marriage, keep her from your wife's flower beds and kitchen garden. Goats are good jumpers, so to contain her you will need a 3½-ft (1-m) fence.

If the free-range area on your own plot is limited or perhaps non-existent you can do what the cottagers do by tethering your nanny on some nearby common land where she will eat just as happily and nutritiously as in your garden. Remember, though, that she will still need attention; you will have to move the tethering peg at least once a day and you must bring her home to her shed in the evening.

In summer, providing she can forage for herself, she will need little, if any, supplementary feeding. In winter or in bad weather, when she has to be kept in her shed, you can feed her on hay, which is nutritious and can be prepared in sufficient quantity from the edges of your own plot or, once again as the cottagers do, from grass verges of country roads or common land. She will eat and thrive on such vegetables as kale, swedes, turnips and lucerne, while a favourite mash, particularly in winter, is a mixture of small pig potatoes, steamed and mashed, dry bread, crushed oats or flaked maize. Most kitchen waste is suitable for your nanny, especially such delicacies as outside cabbage leaves, stale bread and Brussels sprouts stalks split down the middle.

MILKING

Goat's milk has become increasingly popular over the years so that there is now scarcely enough to go round. Not so long ago it was something of an acquired taste, at least among the general public, but now that the therapeutic effects in cases of digestive ailments and skin disorders have been recognized there is always a demand. Put an advertisement in your local paper or get in touch with a health food shop if your nanny produces a surplus: I don't think you will be disappointed. This could be a profitable sideline.

Or you can make your own goat cheese, which is milder and more easily digested than ordinary cheese, and this, too, can be readily sold if you have a surplus.

The truth is that there are so few goats in this country that you will always be in a seller's market. On the Continent it is different. You have probably travelled, as I have, through certain areas of France where every farm seems to bear a notice, *A vendre – fromage pur chèvre*, 'pure goat's cheese for sale'. (I wonder what they do with the cheese of goats that are not so pure: send it to gay Paree, perhaps.)

Goat butter is less successful, at least in my experience, for although it is pleasant enough to spread on bread it is no good for cooking.

The actual physical process of milking is quite simple: you can get the hang of it in one easy lesson, as the advertisements say, and become an expert in a week. Goats should be milked twice a day, preferably once every twelve hours, for example at 7 a.m. and 7 p.m. and – the golden rule of all dairy work – milk cans and utensils must be sterilized after use.

If your nanny is a good milker she will probably produce 100 galls (450 l) a year, although if she is very good she could easily produce twice as much. 3 to 4 pts (1·5 to 2 l) a day is about average.

Now as one who has spent many back-breaking hours in the milking shed, a word of advice. Build a simple milking platform or use a low table so that when you are milking the nanny is standing on your level. You will be surprised how much easier the job becomes.

BREEDING

When you have had your nanny, or nannies, for a time and have found for yourself how easily they can be kept you may feel that you want to breed. Well, this is quite simple.

The female goat comes into season in the autumn and then at three-weekly intervals until early spring, that is approximately September to February. You would be unwise to mate her until she is about eighteen months old and well developed. The gestation period is about five months so that most kids are born in the period of March and April. There are normally one or two kids at birth.

For the first experiment you would be well advised to mate your nanny with a billy goat hired from a reputable local breeder and, so far as possible, you should check up on the billy's pedigree and record.

If this first experiment is a success and you want to continue to breed and, by this time, you have several nannies, it will pay you to keep your own billy. The cheapest way is to buy a year-old buckling, taking care once again to check up on antecedents and record. As with the other farmyard animals the male is so important that it would be wise to have a vet look over your intended purchase. Sexual organs will be one aspect of his inspection to make sure that you are not landed with a goat that is infertile.

When you get him home the billy goat should be housed in separate quarters, for, to put the matter as politely as possible, he tends to be something of a stinker. In fact so intrusive is his smell that it will stay with you long after you have left the goat shed unless you wear a protective smock or apron and it will even taint the milk of nanny goats with whom he has been in contact. The old tag, 'Even your best friends won't tell you' certainly applies to the billy goat.

Towards the end of the gestation period the nanny goat's udders will begin to fill and the nearer she gets to kidding the more they will tighten until she has to stretch herself in a quiet corner for comfort. When she is near kidding a slime will start to come from the vulva.

If she seems to be in pain or has been straining for some time don't hesitate to send for the vet. It may mean that the kid, or kids, are too large to be born without help.

The kid should weigh about 5 to 8 lbs (2·3 to 3·6 kg) at birth and you will have to decide then whether you want to leave it with its mother or to hand rear it with a bottle. Many breeders think that they get better results by drawing the milk from the mother and bottle feeding the kid, either from birth or from the fourth day. This, of course, takes time and needs a lot of your attention.

The secret of successful bottle feeding is little and often. Don't pump as much milk into the kid as it will take unless you want it to be pot-bellied, but ration it to just over a pint (about 0·6 l) of its mother's milk a day for the first week and gradually increase to 4 pts (2 to 3l a day) at four weeks. After that you can start introducing it to a good milk supplement which contains all the necessary trace minerals and vitamins.

On the other hand if you leave the kid with its mother you will have to wait two to three months before it can be weaned, and even six months is not too long.

During the weaning period the time with the mother should be reduced and a mixture of skimmed milk and cod liver oil substituted for natural milk. After this it should not be too long before the kid is able to forage and find its own solid food.

DISEASES

When you have lived with goats for a while you will find, as with other farmyard animals, that you can quickly tell when they are off colour. A sick goat will go off its food – always a danger sign – and there may be a noticeable drop in milk production. The coat gets dry and tends to stand up and the droppings will be soft and runny. The most ominous sign of all is when it begins to lose weight.

Call in a vet at once and isolate the sick animal in a warm dry shed where she can rest in comfort and can receive your special attention. Feed her little and often and in winter fix up an infra-red lamp. Heat, even more than food, works wonders with all sick animals. You will soon know when she is getting better by her improved appetite and her perky manner.

On the other hand if she does not improve and you have to send for the vet don't despair. He may easily have her on her feet in a few days with an antibiotic injection.

Parasites

Many goats are attacked by parasitic worms which they pick up when feeding, especially on overworked pasture. The infected goat will lose weight and condition, but the disease can generally be cured by a dosing with an approved anthelmintic obtainable from a veterinary chemist. Prevention, of course, is better than cure, and you can avoid this troublesome complaint by changing the pasturage fairly frequently and keeping the goat houses clean.

Bloat

This is another common ailment caused by over-eating fresh grass and other green food. The goat's stomach will swell up making it very uncomfortable, although the vet will soon put her right. Kids are particularly susceptible to bloat.

The best preventive is to cut your green food and root crops a day before feeding and to let them thaw out. Wet or frosty green food can be fatal.

Scour

This is also caused by eating too much wet or juicy grass or, in the case of kids, by taking milk too quickly. The infected animals suffer violent diarrhoea and, for a time, feel very sorry for themselves. The white of an egg mixed with water will often work wonders.

Poisoning

We mentioned this before but it is worth repeating since it is a common enough complaint with goats. Since they are such unfussy eaters they tend to find things which may look or smell all right but which in fact are poison. Yew leaves and berries we have mentioned; rhododendron, laurel, laburnum, ivy and box are equally dangerous. Keep an eye open for all these delicacies on your goat plot.

Mastitis

This is an unpleasant condition as it can cause loss of part of the udder unless treated quickly. The udder generally goes hard and develops lumps so that milking is difficult and what little milk you get comes out in clots.

Mastitis is infectious, so isolate the infected nanny at once and disinfect everywhere she has been. Make sure, too, that your hands are disinfected.

The vet should be called and, if the disease has been caught at an early stage, treatment with an antibiotic will usually effect a cure. Delay can be fatal.

Lice

I suppose we should mention these unpleasant creatures, which are common enough to most farmyard animals. Cleanliness is the best preventive, but if your goats still scratch themselves raw and look uncomfortable a good dusting of insect powder or flowers of sulphur will usually do the trick.

THE POOR MAN'S COW

In some parts of the country, especially among the well-to-do farming community, the goat is sometimes known as the poor man's cow. Well, if that term is derogatory so be it, but if I were a small-holder or a backyard farmer I would not dream of being without one. Your nanny goat is hardy and inexpensive and virtually trouble free. She will keep you supplied with excellent milk and cheese, she will produce one or two, or sometimes three or four, kids a year: and all this on the roughest of pasturage, supplemented only in winter and bad weather with such hay as you can cut from common grazing land or from your own plot. She will eat the scraps from your table and the waste greenstuff from your kitchen garden. Poor man's cow, indeed: she is a farmyard treasure!

9   Sheep

To those who feel too inexperienced to cope with pigs or perhaps have neighbours who protest abhorrence of all things porcine may I recommend that modest and gentle member of the farmyard community, the sheep. Apart from its baa-ing, which can be anything from soothing to deadly monotonous, according to your taste, there is hardly a blemish of nature, appearance or even smell which precludes this modest quadruped from being a good neighbour. Enclose her in a pasture surrounded by a good wattle fence and she will give you no trouble, indeed you may find, as will your neighbour, that there is something soothing, perhaps even soporific, in her placid munching.

So, first things first: let us consider your plot. An acre (0·4 hectare) of land will support about four ewes plus their lambs, but if part or most is pasture and you have a shed suitable for wintering it may support a few more.

Not that you will want to start on that scale. One or two ewes at first, with more coming along from breeding is sensible. It is far better to start modestly and arrive than to spoil everything with an impetuous dash. Experience will bring knowledge and from knowledge springs confidence: after one or two instructive and profitable seasons you will know how much you can manage.

BREEDS

Britain is one of the leading sheep-producing countries in the world, and although not so important as in the Middle Ages when wool was our most lucrative export, sheep farming is still a major industry. British breeds are much sought after, and the export of attested thoroughbreds is still an important part of our economy.

Breeds are many and varied but can be conveniently classified under three headings according to their appearance and, more important, the terrain of origin. Thus we have the Mountain breeds developed in the highlands and moorlands, the Shortwools, including the Down breeds, coming mainly from the south and west of England, and the Longwools, which are found mostly in the Midlands.

The Mountain breeds, which include Cheviots, South Blackface, Welsh Mountain, Exmoor Horn and many others, are hardy little animals, used to grazing on heather and scrubby grassland and to being exposed to all kinds of weather. They are bred mainly for their mutton, which is tender and succulent.

The Shortwools include Hampshire Down, Southdown, Oxford Down, Dorset Down as well as Shropshire, Suffolk and other familiar breeds. The Southdown, which is the smallest, produces excellent lamb as well as the most sought after wool.

The Suffolk, which resulted from a cross between the Southdown and Norfolk Horn, is probably the most popular breed in Britain.

The Longwools, which include Leicester, Cotswold, South Devon and Wensleydale, have long, fairly coarse wool and are all white-faced.

So, you may ask, which breed would I recommend?

Well, coming from the west country I am naturally partial to the Devon Closewool and the Devon Longwool, but your best bet would probably be a cross, say a Down ram on a crossbred, which would give a good meaty sheep, full of hybrid vigour. But having said this I must qualify it by saying that your choice would be much better guided by the breeds which are farmed locally. If you find that the Southdown or the Suffolk are popular in your area you could hardly do better, but if not then your safest bet would be to follow your neighbours. Buy a ewe in lamb or a ewe with one shearing, called a theave (thave, your farming friends will call it).

When you get her home fence in a fairly small area of paddock so that you can move her quite frequently from one patch of pasture to another. 'Sheep don't like to hear church bells twice in same pasture', is an old country saying, and, like most of these sayings, it contains a deal of truth. Sheep thrive by frequent moving. While the pasture is good they are the most placid of creatures, but once the grass becomes short they become determined and even aggressive in their search for pastures new. You have probably seen many a stray sheep in country lanes, sheep which have forced their way through substantial hedges, lured on by the thought that the grass is always greener on the other side of the fence. Try to avoid this situation if you can, for a confused and bewildered sheep is a hazard on the roads, dangerous to motorists as well as to itself.

If, on the other hand, you find that your capital is low and you feel that for the moment a sheep is beyond your means, get talking to your neighbours and see if any local sheep farmer would be interested in half sharing.

It frequently happens that a successful breeding farmer finds that he has too many sheep for his land. Sooner than send the surplus sheep to market he prefers to place some of his flock with another farmer with less capital on the understanding that he will take half the gross profit.

Such an arrangement might well suit you. The sheep remain the property of the breeding farmer, who will take half the lambs and may want half the wool, although many farmers are happy to leave all the wool with the foster farmer.

Another way to start is to buy a few wethers, that is, lambs which have been castrated, and run them for a few weeks or months on good pasture or, if you have a vegetable garden, on a diet of hay and concentrate plus any swedes, turnips, flat pole cabbages or kale you can spare. Fat lambs ready for the butcher fetch a good price.

Yet another way of starting is to buy a draft ewe, that is one which is sold from the flock after three or four shearings. She will generally not be of top quality, partly because of her age and partly because her incisor teeth may have been worn down by rough grazing. Such ewes, broken mouths we call them, which can be bought relatively cheaply, will respond to better pastures and can be quite profitable. They are usually bought in late summer or autumn, put to the ram, and the lambs are fattened off and sold the following spring. With proper care and attention a draft ewe will produce one or two more crops of lambs.

During the summer months a sheep will need little, if any, extra feeding and, since it is a hardy creature, will live quite happily in your paddock. In winter or in prolonged spells of bad weather it should have access to a shed or, at least, to a corrugated iron lean-to lined with straw. With their thick coats of wool they are relatively well endowed against bad weather but unless you have a mountain breed they will need some extra protection.

Their natural diet, of course, is grass and other greenstuff, so your paddock or half-acre (0·2-hectare) plot will do very well. Don't be alarmed if your newly acquired ewe seems to make short work of your pasture. She is by nature a heavy eater, but if you move her, as I suggest, from one patch to another you can, with a top dressing of nitrogen fertilizer, so improve the growth on the spent patches that she will always be moving to fresh green grass.

Incidentally, if you live near moorland it would be worth inquiring about common grazing rights, although moorland farming is really outside the scope of this book. To make it profitable you would need a fair-sized flock and, to be frank, more knowledge and sheep lore than you would be likely to have as a beginner, but it is worth thinking about for the future.

BREEDING

So let us return to my original suggestion and consider the ewe in lamb and the alternative suggestion of a theave.

Breeding must be your object, for although wool will bring you a respectable profit, mutton is where the money lies.

Take first the ewe in lamb. She will come to you some time in the autumn or early winter and the lambing can be expected roughly twenty-one weeks after mating, that is from January onwards in the south, any time up to April in the north. She will need warm dry quarters during the gestation period and for some little time afterwards, so you will need a shed or, better still, a barn. Remember that she is an expectant mother

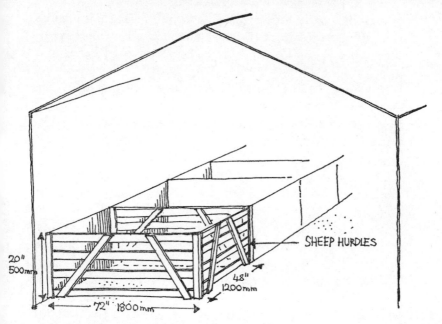

SHEEP HURDLES

20"
500mm

48"
1200mm

72" 1800mm

Individual pens for ewes and lambs

and although she doesn't need coddling she must be kept quiet, warm and well fed.

The alternative to a shed is a lambing pen which can be quite easily made with a few sheep hurdles driven into the ground to form a square, plenty of straw against the hurdles and on the ground to keep out cold and damp, and a corrugated iron roof. It won't be as comfortable as a shed, at least in my opinion, but plenty of lambs are brought into the world in this way.

Whether you use a barn or a shed or a pen be sure to include a hay rack and a trough containing a ration of ewe and lamb nuts plus a few chopped turnips, swedes or flat pole cabbage.

A few weeks before lambing the ewe should be crutched, which means that you remove the dirt from around and under her tail with a pair of clippers.

From the end of the twentieth week onwards you will have to keep an eye on her to make sure that you don't miss the lambing. All animals, including farmyard stock, like to get into a quiet corner for the actual birth and, since this is nature's way, it is best to accept it. Don't interfere unless you have to and, whatever you do, don't let her be disturbed by extraneous noises.

Having said this I must add that there are times when the shepherd's help is necessary. The lamb should be born head and front feet first; if, as sometimes happens, it has turned backwards in the womb you will either have to correct it yourself or send for a farming neighbour or vet.

The ewe will have one, two or even three lambs, and I think that when you see them for the first time you will get quite a thrill. They are such long-legged, helpless little creatures that you can't help feeling sorry for them, especially when they stagger about the straw searching for the food they know is there.

However many lambs are born you will feel pleased, but if there is only one don't be disappointed for many shepherds say that it is better to have one strong one than two littl'uns and that trebles, or triplets, can be a positive nuisance. Their reasoning, based on experience, is that because the ewe only has two teats one at least of the offspring is likely to get short rations.

In fact if your ewe has trebles you will have to watch her carefully and, from time to time, take away one lamb, the strongest, so that the third, the weakest, can feed. The mother won't like it and will probably get quite agitated when she hears the plaintive baa-ing, but remember that each lamb is an asset and if you can rear three, rather than two, so much the better.

To be honest I must confess that the female sheep is not the brightest of creatures. I have had ewes bearing two lambs, the first easily enough, the second with difficulty, and have found that in the time between the births she will, as like as not, have forgotten the first born and only accept it with difficulty. Similarly with trebles she will show real concern for the two lambs she is suckling but no concern or even interest for the unfortunate third.

In fact, if you have several ewes lambing at the same time your best bet will be to take the third born from its own mother and put it to a ewe which has only one or, better still, which has lost her own.

This isn't as easy as it sounds. Although ewes are not exactly bright they know enough to recognize and reject a lamb which is not their own. They do this by smell, and you will probably have seen in a field full of sheep and lambs how some lambs, having lost their mothers, will rush from one ewe to another trying to nuzzle, only to receive one butt after another for their pains. The ewe will only accept her own.

So when you are faced with this problem, as you undoubtedly will be when you increase your stock, you will have to use a certain subterfuge. It is unfortunately true that however much you protect your ewes one or more will have her lamb born dead. When this happens, and with as little delay as possible, you should strip the coat off the dead lamb and tie it on the third born of a treble. Don't worry what it looks like, the ewe depends entirely on smell. Introduce the foster lamb to her gently, and soon, perhaps after some initial hesitation, she will accept it. If after several attempts she still seems reluctant bring your dog into the shed or pen on a lead. It seems a mean trick, but the mothering instinct is so strong that she will generally rush to protect the foster lamb.

If you have to introduce the third born to a ewe with one live lamb of her own your task will be much more difficult. No trick is likely to help here: you will only succeed by patience. Wait until her own lamb is suckling before introducing the stranger to her spare teat. She won't like it and will almost certainly reject it at first, but keep trying. Eventually she will get used to it and after that you should have no trouble.

Incidentally if you or your wife feel so sorry for the little outcast that you would like to hand rear it, this is quite possible – I have done it myself, many times – but it needs time and patience, and, the trouble is, you get so fond of the little fellow that you find it almost impossible when the time comes to send it to market.

Each lamb should weigh about 10 to 12 lbs (4·5 to 5·4 kg) at birth, although one born of a mountain breed on a hillside farm may only weigh half as much.

Now, in case you think I have forgotten, let us go back a

few months to the theave, that is, a ewe after her first shearing. She will be ready for mating so that her lambs will be born when she is about two years old. As it would be plainly impracticable to keep your own ram, at least until you expand considerably and have twenty or more ewes to serve, you will have to hire a ram from a local breeder.

Remembering that the gestation period is about twenty-one weeks, you will want to time the mating so that the lambs are born in early spring. In this way they will arrive when there is a plentiful supply of natural food, they will have all spring and summer to fatten, and the ewe will be ready for a second mating in the autumn.

Oestrum lasts for twenty-four hours and if the ewe has not conceived is repeated in sixteen days. The mating time covers three oestrum periods.

After mating the procedure is exactly as I have indicated above.

Weaning

The lambs stay with their mothers until they are three to four months old. During this period they will need little attention, for they are lively little creatures and generally remain healthy. Most farmers separate lambs from ewes at shearing time in June although others prefer to leave them together for the full four months.

Castration and docking

Male lambs not intended for breeding should be castrated at as early an age as possible and all lambs should be 'docked' or 'tailed'. Sheep which have long tails are difficult to manage because the wool becomes soiled and stained around the tail.

Castration and docking may be carried out by the stock owner using an elastrator and rubber ring. The rings are placed over the scrotum on to the cords, just above the testicles. Similarly with tailing the ring is placed about 1 in (2·5 cm) from the top of the tail, thus leaving a slight 'dock'.

The best thing for a beginner is to purchase an elastrator and box of rings from a veterinary chemist and then ask a professional shepherd or veterinary surgeon to show you how to apply the rings.

Lambs are normally rung when one to two days old, and thus the minimum amount of discomfort is caused.

DIPPING

Ministry of Agriculture regulations require you to have your sheep dipped once a year, which is an excellent precaution against the disease known as sheep scab and all sorts of parasites. Until you expand your farm it would be an unnecessary expense to prepare your own dip, so I suggest you get in touch with one of the local farmers and ask if you can run your sheep with his when he is using the dip.

SHEARING

Although wool will be only a sideline profit, at least until you get more sheep, it is not to be despised, in fact you may be pleasantly surprised after your first shearing. Get an expert contract shearer to help you.

June is the month for shearing, when the wool has become greasy, which, incidentally, is nature's way of saying that so much wool is superfluous. Keep an eye on the weather, though, for a shorn sheep or lamb can quickly catch cold. If it turns cold or wet you would do well to take your little flock into shelter for the night.

WINTERING

And talking of shelter don't forget that your sheep should be kept in warm dry quarters in winter. You may ask whether this is essential, for you will have seen pictures or read in your newspapers of sheep being rescued or having bales of hay thrown to them from helicopters during severe winter conditions, and it is true that most sheep, particularly on the warmer lowlands, will come to no great harm if they are left outside. You may even have seen, as I have, ewes with lambs huddled against them for protection against the driving snow, but no farmer, I think, would choose to subject his flock to such conditions or risk the inevitable losses. Lambs, remember, are your future capital. They must be protected.

No, the fact is that you are in a rather different position from the big farmer. The ewe represents a fair part of your capital and the lamb, or lambs, she is expecting represent future profit. The big farmer has a shepherd or stockman to keep an eye on his flock and, when conditions get really bad, he will either drive it to the protection of a barn or possibly rig up a makeshift shelter with bales of straw and corrugated iron.

The disadvantage of keeping your sheep under cover, of course, is that they will need feeding, but at the end of the day when they arrive at the spring lambing fat and in good condition you will find that you are little, if anything, out of pocket. And remember that the costs you incur in winter feeding will be partly offset by the much improved crop of grass you will have next summer.

Your ewe will need 8 to 12 ozs (225 to 350 g) a day of concentrates, which you can buy in cube form, plus hay and root crops, such as swede and turnip and any garden greenstuff such as kale or cabbage. If, as I suggest, you have a half-acre (0·2-hectare) kitchen garden, you can probably grow most of this yourself.

COSTING

Now remembering the provisos we made in the chapter on pigs it might be helpful to show how you could prepare another costing. The actual figures would have to be inserted by yourself using prices current at the time.

A good ewe will cost say £x, a top class one a good deal more. The saleable product, apart from the wool, is the lamb.

You find that lambs are currently selling at say y pence a pound dressed carcase weight. The demand is for a 40 to 45 pound lamb with a minimum of fat, so a good 42 lb lamb will sell at approximately $£\dfrac{y}{100} \times 42$.

Assuming a ewe produces one and a half lambs there may not be much margin of profit remembering that she eats approximately 1 cwt concentrates and 2 to 3 cwts of hay (which can also be costed) in winter.

On the other hand she could stay with you for up to four years, and four years' lamb production from one ewe makes the operation worth while.

DISEASES

There are numerous diseases which can affect sheep, but I am only going to mention one, foot rot. Many of the others – tetanus and the like – can be prevented by a vaccination programme.

The best thing to do is to talk to your vet and take his advice about prevention of disease. He will also supply the necessary vaccines and medicines.

Foot rot

This is one of the most prevalent diseases among sheep and is one that can be cured without too much trouble provided you catch it early enough. Most farmers inspect their sheep when they are dipped, but with only one, or possibly two, in your flock you should be able to make more frequent inspections.

Foot rot is a painful and contagious disease which will quickly render your sheep lame in one or more feet. Infection enters between the 'toes' or in any place where a hoof becomes separated from its surrounding tissue. Inflamation sets in and a troublesome sore develops. Damp or soft ground allows the horn to overlap, and infection frequently begins in these folds.

Since prevention is always better than cure it will pay you to inspect the hooves regularly and pare away any overgrown hoof, using a sharp knife or secateurs. You may then dip the foot in a solution of either 10% copper sulphate or 6% formalin.

If, despite these precautions, they still go lame then consult your vet.

D

I hope that having read this chapter you will feel encouraged to try sheep. Although they are not the brainiest of animals they are some of the most docile and if properly looked after the most rewarding. They can be annoying at times, as when you want to herd them from one enclosure to another, but they are patient and forbearing and, on the whole, remarkably free from disease.

Their young, of course, are entrancing and, no matter how long you farm, I don't think you will ever lose the thrill of seeing new born lambs.

To keep sheep properly you want a small-holding although a really large, well grassed garden could support one or two ewes. If you have the room, and the enthusiasm, have a go: I don't think you will regret it.

Whatever else you keep in your backyard you must, I repeat must, keep rabbits. Why? Because they are the most prolific, trouble-free and profitable farmyard animal of all.

Think of it this way. With some simple hutches, which you can easily make yourself, some straw or sawdust for bedding and a minimum outlay for food a rabbit family of, say, four does and a buck will supply you with at least one rabbit dinner a week. More rabbits mean more meat for your table and more money in the bank, for you should have no difficulty in disposing of any surplus stock. Could anything be more simple?

You will appreciate, I am sure, that I am talking about the common or garden rabbit, first cousin of the little fellows you see in a field. There are, of course, aristocratic offshoots of the family whose members are kept mainly for show, such as the Harlequin, the Lop, the Belted Dutch and the Belgian Hare, but these are outside the scope of this book.

Two breeds I can recommend are the New Zealand White and the Californian, which can be bought quite cheaply (about £6 a head at the time of writing). They are hardy, easy to keep and quiet to handle and should each weigh about 12 lbs (5·5 kg).

The Giant Flemish, another favourite, is the largest of the breeding rabbits generally found in this country, the fully grown buck weighing up to 15 lbs (6·7 kg). However, you can do perfectly well with one of the mongrel breeds generally on sale in market places providing the animals are young and healthy. Remember, though, that size is important, for you cannot expect a small doe to produce big rabbits.

HOW TO START

First, as with all livestock, you will need a costing. With feeding stuffs at their present high level this is just common sense, but I shall be surprised if you do not decide that rabbits are worth while. Having said this, one word of warning in case the venture seems completely fool-proof: rabbit farms do fail, usually because of disease. This, of course, is a risk you must take, but provided you give your rabbits plenty of attention and, above all, keep them clean, this risk is minimized.

I suggest you start with a buck and four does and build up your stock by breeding. Gestation period is thirty-one days, and a healthy doe can produce at least three litters a year. As the average litter is five or six (although there can be as many as ten) you will see that even this small colony will multiply rapidly.

HOUSING

There are only two types of housing you need consider. The first, known as the Morant, consists of a hutch with raised floor, lined with straw, and a run similar to a chicken fold. The only difference is that the wire netting also covers the floor of the run, for remember that rabbits are natural burrowers, and your stock will feed on the grass that grows through the meshes.

This kind of hutch has many advantages and helps keep the rabbits in good health. It can be moved from one grass patch to another, just like a chicken fold.

The other method, which may be more practical as your stock begins to multiply, is a series of hutches placed in a shed or barn or kept outside if they are properly weatherproofed. The usual plan, which you would do well to copy, is a nest of hutches arranged in tiers of three.

To start with you might construct a unit of six hutches as in the diagram facing, although as your rabbitry expands you

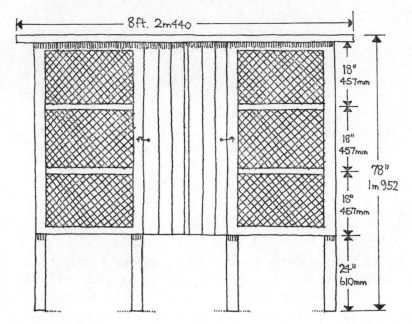

Six hutch unit

may have to think of larger units extending along the walls and then across the floor of your barn or shed.

The plan you adopt must depend to a large extent on the accommodation available, but whatever you build should comply with certain basic measurements.

Each hutch should be approximately 4 ft (122 cm) long, 2 ft 6 ins to 3 ft (76 to 91 cm) wide, and 18 ins (45 cm) high. Unless you are going to build separate nesting boxes, one third of this area should be partitioned so that the doe can keep her babies warm. (The alternative is to build separate nesting boxes on exactly the same lines but only half the size. Personally I find it easier to have all my hutches of a uniform shape and size.)

Another tip, which is for your benefit rather than the rabbits', is to build the tiers so that the bottom of the lowest hutch is 2 ft (60 cm) above the ground. This is not only a protection against damp and vermin but enables you to clean the hutches, as you will have to do every day, without breaking your back.

HANDLING

One point is worth a paragraph to itself: handling. Don't, I implore you, pick up your rabbits by the ears. You probably remember the furore a certain President of the USA aroused when he picked up his dog in this manner. Well, a rabbit is just as sensitive. The correct method is with one hand grasping the ears, the other supporting the body by the haunches. It's not only a question of kindness, but common sense.

FEEDING

Food should be no great problem, for the rabbit, being herbivorous, will exist quite happily on the greenstuff and vegetables you can collect from the hedgerow and garden. It is especially partial to grass, cabbage leaves, clover, chickweed, groundsel, lettuce and dandelion. In winter you can substitute root crops such as swedes, turnips and mangolds plus hay, of which it is especially fond, and a mash made from kitchen waste, cooked small potatoes and bran. During pregnancy the doe should also be given some rabbit pellets, which can be bought at any corn chandlers or pet shop and which contain all the ingredients to keep her healthy.

Two points to remember: first, never give your rabbits grass or vegetables which are wet or have frost on them, as this is a sure way of making them ill, and, second, beware of certain hedgerow plants which are poisonous e.g. bluebell, buttercup, celandine, foxglove, hemlock, henbane, poppy, night-shade and the leaves of most evergreen shrubs. Acacia and laburnum are also poisonous.

Rabbits should be fed twice a day, the daily diet in winter for an average size breed being 8 ozs (225 g) of greenstuff, a large handful of mash and 3 to 4 ozs (75 to 100 g) of hay. In summer the diet will consist entirely of greenstuff.

I have said before, but I will say it again, that cleanliness is essential. Uneaten food should be removed before it attracts vermin, and the cages should be thoroughly cleaned daily.

Incidentally, the rabbit droppings you collect will be excellent for the garden.

Lastly, each hutch should have a supply of fresh water. You can buy, or improvise for yourself, small automatic drinkers, with glass or metal feeds so that the rabbit can't gnaw the end. In this way they will have a constant supply of fresh water.

BREEDING

Rabbits are first mated when they are about eight months old, and a young buck, in prime condition, can service three to six does a week.

When mating always bring the doe to the buck otherwise you may have a fight on your hands. In normal mating there is a preliminary stamping of feet followed by a pause before the buck mounts the doe. If the doe is not ready for mating the buck will fall off in a few minutes and the buck should be removed. You can reckon that if the doe does not mate in the first five minutes she is not ready.

Three weeks after a successful mating the doe will start gathering straw to make a nest, so be sure that some is available. She will also pluck fur from her breast to keep the babies warm.

The actual birth, called kindling, takes place about a month from the mating. The babies, which will usually number five or six, although litters can vary from one to ten, are blind and almost naked, although the fur grows rapidly and the eyes open in nine to ten days. If you have several does kindling at the same time you can remove a baby from one doe to another if the size of litters varies considerably. Unlike the sheep, the doe does not seem to mind suckling a stranger.

During these first days the doe should be disturbed as little as possible, especially if she is a maiden doe, that is, with her first litter, for she will take fright easily and when frightened will often kill her young.

In two weeks, or thereabouts, the young rabbits will be able to scamper about the hutch and a week later they will start nibbling some of their mother's solid food: at eight weeks they should be weaned.

To do this it is best to remove the doe to another hutch leaving the youngsters in the familiar hutch where they were born.

From then onwards they will grow rapidly, especially if they are well fed, and should be ready for the table when they are three to six months old.

MARKETING

Some of the table rabbits you will need for your own larder or deep freeze: the surplus will find a ready market. Contact your local poulterer who will probably be glad to buy them by live weight, thus saving you the actual killing.

Rabbits being used for meat should not be fed the day before killing. A poulterer or rabbit dealer will show you how to kill your table rabbit by a simple method of dislocating the neck.

Skin the rabbit at once and then, after cutting the throat, hang it head downwards so that the flesh drains white.

Apart from local poulterers and butchers you would do well to contact hotels, especially if you live near a seaside or holiday town. The hotelier will probably be glad to take all your freshly killed rabbits and may even get you to provide a regular supply by contract. This can be a very profitable arrangement.

You can also make money from the pelts, which, as you probably know, form the basis of many fur coats.

After skinning the rabbit nail the pelt to a piece of wood or pelt stretcher, which you can buy from a commercial rabbit supplier. It must be dried and stretched and any surplus flesh or fat removed from the skin, taking great care not to damage the pelt. When completely clean the drying should be completed outside in the air, but not in the sun and never before a fire. The skin will take about three weeks to dry.

If you want to store the pelts until you have enough to offer for sale make sure that they are properly dried out before you pack them. Place them in a flat box, fur to fur and make sure that no mice or rats can get at them.

Commercial buyers will always take your pelts, but you

Above. A Large White gilt.

Below. A tethered Large White sow feeding her piglets.

Above. The Aylesbury Duck is hardy, docile and a quick grower, although its egg yield is low.

Below. A Broad Breasted Bronze Stag Turkey with a turkey hen.

Above. The goose is the first recorded domestic bird. One of today's most popular breeds, seen here, is the Toulouse.

Below. The Indian Runner and the Khaki Campbell (the smaller birds) are breeds of duck which can be relied upon to produce large quantities of eggs.

By full blossom time in May the hive is a crawling mass of bees. When the hive becomes too overcrowded the bees swarm.

will probably get a better price if you can find someone locally who makes fur animal toys or novelties for the holiday trade. Look in the Yellow Pages of your telephone directory.

DISEASES

If you keep your hutches clean you should have little trouble with illness. Bites and sores and even a broken limb are more likely to trouble you, for rabbits kept together in enclosed spaces tend to fight. However, you should soon find the trouble-maker and remove him to a separate cage.

Coccidiosis

This is the most serious and prevalent disease of rabbits. It is caused by a small parasite, and the infected animal usually suffers from diarrhoea, loses weight and huddles in a corner. You may find several of your colony suffering the same symptoms; if so, take them to a vet at once and he will probably cure them with antibiotics.

Worms

These can be tape worms, which will be detected as long worms, 1 to 10 ins (2·5 to 25 cm) in length, in the excreta, or roundworms, which are half an inch (1 cm) in length and coloured light red. The vet will supply the infected rabbits with worming tablets.

Bloat

This usually occurs in the spring due to overfeeding on succulent grasses and clover. The rabbit is blown up by the amount of gas in the abdomen and is in great discomfort. The disease is usually fatal.

Let me end by repeating what I said at the beginning of this chapter: for the backyard farmer a small colony of rabbits is a 'must'. The total investment is small, the cost of upkeep negligible and the profits are virtually assured.

Bees

If you have never kept bees before and have not read any of the authoritative manuals such as Maeterlinck's *The Life of the Bee* you are in for a treat, for no creature, certainly no domestic creature, is so full of interest. You probably know that bees help to germinate blossom and are therefore good for the orchard; you probably know that bees have queens and workers and drones and that they tend to swarm in June. You probably remember the old country saying,

> A swarm of bees in May is worth a load of hay;
> A swarm of bees in June is worth a silver spoon;
> A swarm of bees in July is not worth a fly.

But did you know that the colony, or stock as it is more commonly called, is run as efficiently and ruthlessly as any totalitarian state?

To explain what bee-keeping means to the backyard farmer it will be best if I tell you what life is like in the hive.

As a start let us consider the population. There are one queen, up to 1000 drones and, in early summer, perhaps 50000 workers or more. Queen and workers are female, drones are male, but before you start talking about women's lib let me tell you that every autumn these unemancipated females drive out all the males to starve and die.

The queen is regal in every sense – she is the mother of the colony. When she leaves, the colony, or a good part of it, follows. She is easily distinguishable by her shape, which is long and pointed with a pronounced sting in the tail. Her job, in fact her only job, is to lay eggs which she does in season to the extent of 2000 to 3000 a day. The queen cell is reared in sixteen days and she may live as long as five years.

Drones, the males, have no sting and their only job is to fertilize the new queens in the spring and enjoy life while they may. The drone cell is reared in twenty-four days.

Everything else in the hive is done by the workers, whose life, short and hectic in the summer but more sustained in the

QUEEN DRONE WORKER

Bees

winter, is ceaseless effort. A woman's work is never done, they say: well, this is certainly true of the bee.

The worker cell is reared in twenty-one days and from then on there is no let-up. For the first fortnight of their lives the workers stay inside the hive, feeding the queen and the young grubs, cleaning the cells, fighting off marauding robber bees and wasps, building up the combs and, in case they have a spare moment, fanning their wings to keep the hive cool. When they are two weeks old they leave the hive to begin their work proper.

This consists of a constant to-ing and fro-ing, collecting water, bee gum, nectar and pollen. You have probably seen worker bees alighting on the landing board and crowding busily into the entrance of the hive, the pollen baskets on their back legs full and the wax pockets beneath their stomachs giving them an oddly pregnant appearance, although, except in special circumstances, they do not lay eggs.

You may not be surprised to learn that, worn out by this harsh regimen, their life span is only eight weeks, although in winter, when there is no pollen or nectar to collect, they may survive for eight months.

THE QUEEN

The queen, around whom the whole colony operates, starts laying eggs towards the end of January or early February according to the weather. She lays two kinds of egg, one which develops into the drone, the other into the female, either queen or worker.

It is important to recognize the queen egg, which is laid in a cell resembling a hanging peanut. Drone (large) and worker (small) cells are hexagonal.

As the spring and early summer flowers appear the egg-laying increases and by full blossom time in May the hive is a crawling mass of bees. They depart and return in their thousands, laden with nectar and pollen and water to feed the young grubs, and thousands more are being born daily.

It is at this time, from mid-May to the end of July, that the hive becomes so overcrowded that the bees swarm.

SWARMING

As you can imagine in such a highly organized community the decision to swarm is not taken lightly. Before leaving, the queen and her followers eat enough honey to keep themselves going for several days. They also ensure succession by leaving several queen cells in varying stages of development. When the most advanced queen cell reaches the end of the larval stage, that is, when the developing queen is nine days old, the old queen and several thousands of her followers leave the hive.

Normally they will make for a hollow tree or an overlapping roof, but if no suitable home can be found they cluster on a branch while their scouts go out to reconnoitre. It is at this time, as we shall see later, that the bee-keeper has to move quickly.

Meanwhile, in the hive, the new queen is still developing. When she is born her first act, unless she is stopped by the workers who have stayed behind, will be to sting to death the other queen cells, thus establishing undisputed succession. However, if the colony of bees still left is large enough the other queen cells, or at least some of them, will be allowed to develop, and in a short time the senior queen will depart with a second swarm, called a cast. Following a really good spring and early summer there might be two or even three casts leaving the hive.

But the bee-keeper need not despair. Even if he has lost the swarm and casts (and he will try to prevent this if possible) the remaining queen will soon mate and the whole busy process of production will continue.

AUTUMN AND WINTER

As the year progresses and there are fewer flowers from which to gather nectar the queen lays fewer eggs and the population of her colony diminishes. During the autumn the drones are driven from the hive and, since the workers, who have been dying off steadily throughout the summer, are not replaced the hive population will be much reduced, reduced in fact to the number which can be sustained until next spring on the honey and pollen already gathered.

Towards the end of January or early February the whole cycle begins again.

THE HIVE

The hive is generally built to a standard design which has evolved over the centuries and is considered to be the most efficient and convenient for managing the stock and producing the maximum amount of honey.

Briefly, it is a wooden construction, built rather like a two- or three-storey dolls' house, each storey being separate and removable. Fitted together they form a solid, weather-proof home, with compartments for breeding and storing honey, which can be easily taken to bits and reassembled.

On a solid platform, raised 10 ins (25 cm) or so from the ground, the lower compartment on ground-floor level consists of a brood chamber, while above this, separated by a sheet of zinc with holes too small for the queen to squeeze through (the queen excluder), are what we call 'supers' in which surplus

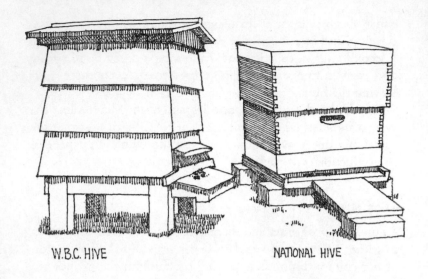

W.B.C. HIVE   NATIONAL HIVE

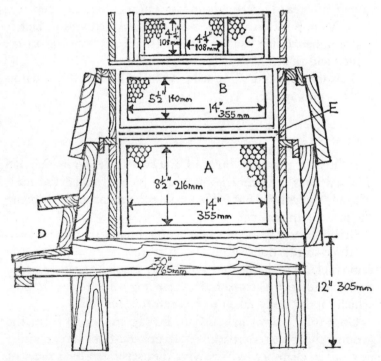

$4\frac{1}{4}$ 108mm $4\frac{1}{4}$ 108mm   C

$5\frac{1}{2}$ 140mm   B   14" 355mm   E

$8\frac{1}{2}$" 216mm   A   14" 355mm

D   30" 765mm

12" 305mm

Beehives

honey is stored and from which it is later extracted. The top floor is for the storing of comb honey.

The diagrams opposite shows you the lay-out.

The brood chamber is, of course, the most important part of the beehive, for it is here that the bees live, the queen lays her eggs and the grubs develop. Suspended side by side in this chamber and set, from centre to centre, about $1\frac{1}{2}$ ins (4 cm) apart, are brood combs, each 14 ins (35·5 cm) long and $8\frac{1}{2}$ ins (21·5 cm) deep. A space of $\frac{1}{4}$ in (0·6 cm) must be allowed between each end of the comb and the outside wall to allow passage of the bees from one comb to another.

The actual combs are built by the bees on wooden frames fitted with sheets of wax which you, as bee-keeper, must supply.

Above the brood chamber is a sheet of zinc, the queen excluder, already described, and the supers, from which you will draw your supply of honey.

Usually the chamber immediately above the brood chamber contains long, shallow supers and the honey when extracted has to be separated by rotation in a simple device called a honey extractor. The top chamber contains the familiar square frames of comb honey you buy in a shop.

HOW TO START

If you have no experience you would be well advised to start with a nucleus, that is a small stock left after the main stock has swarmed. You will find only five or six brood combs, as against the usual ten or more, but provided you are able to start early enough, say before the end of June, you should get a small surplus of honey in your first year.

The big advantage of the nucleus is that it enables you to gain experience while your stock grows. In the second year you should expect worthwhile returns.

Alternatively, if you are impatient or have had experience, you can buy a full stock on their combs, and with reasonable luck, especially with the weather, and efficient management you should get a reasonable surplus in the first year.

Another alternative, of course, is to catch a swarm, which

is done quite easily once you have located it. Your best bet is to pass the word round among farming and gardening friends in the vicinity that you are interested in bee-keeping, and if one of them discovers a swarm in his garden he will probably be only too glad to get rid of it.

The actual removal is quite simple. All you need is a box and a soft haired brush or broom. Bees in swarm are generally not aggressive and you should have no difficulty in brushing them, or shaking them if the branch is high, into a box, which should then be carried quietly and without too much shaking to a vacant hive.

Set your box on the ground, open side downwards, with one edge raised an inch or so on a stone. In the evening shake the bees gently on to a sloping board leading up to the hive or direct into the brood chamber.

Since the swarm will have to start from scratch as it were, building up on the frames you have inserted, you should only expect a small surplus of honey in the first year.

EXTRACTING THE HONEY

Your harvest, or profit, will be the surplus honey you extract during the summer and autumn from the supers placed above the brood chamber. Depending on weather and the amount of flowers and blossom a full stock should provide you with 30 to 40 lbs (13·5 to 18 kg) of surplus honey a year, although in certain parts of the country – near heather moorland, for instance – the yield may be much more.

Extracting the honey is relatively simple once you have overcome your initial fear, and although you can hardly expect the bees to be overjoyed at losing their winter reserve you can, provided you take certain precautions, manage without a single sting.

The first point to remember on this or any other occasion is that bees respond angrily when their hive is knocked, so try to move quietly and slowly and, above all, don't start any panic swatting. If a bee settles on your arm the chances are that it has no intention of stinging. Like other creatures, they can sense fear, so remain calm if you can. If you have a friend who is an experienced bee-keeper watch how he goes about it. You will notice that he uses quiet, confident movements and seems to take no notice of any bees that land inquiringly on his arm. The fact is that they accept him and so long as he doesn't trip over and upset the hive it is most unlikely that they will sting.

Nevertheless, it is wise to take certain precautions.

First of all clothing. For some reason bees hate dark tweedy clothes, and you will be much safer in something smooth and white. A long white smock is ideal.

Your head and face should be covered with a hat over which is draped protective veiling. The veiling, which should have a mesh that you can see through, should be tucked firmly inside coat or shirt. Don't leave any vulnerable spaces in this area otherwise you may soon be hopping mad with an angry bee or two next to your skin.

Cover your arms, if you feel safer, and tie the wrists with string or elastic bands. Wear gloves too if you like although they make for clumsy handling, and I shall be surprised if with experience you don't decide that the white smock and the hat with veil are enough.

Now you need a smoke gun. This is a simple device with a combustion chamber in which cardboard or sacking are burnt, a nozzle and bellows. As you squeeze the smoke into the hive entrance and across the top of the combs the bees become frightened and drowsy and less inclined to sting.

You will also need a hive tool, which is like a blunt, broad-ended chisel and is useful for prising the combs apart.

HIVE MAINTENANCE

A hive of bees will take very little of your time, and, for this reason alone, it is a good investment.

Once a week during the summer months you should examine the combs and destroy any queen cells which, as we saw earlier, are easily distinguishable. Make sure always that the brood chamber and supers have enough combs for the bees to work on.

During the summer, as the combs in the supers become full, you will be able to extract your honey.

E

In the autumn you will have to make sure that the hive has enough food for the winter. 30 to 40 lbs (13·5 to 18 kg) are normally required to keep a full stock going to next spring. As you will already have taken much of their reserve you will have to top up with sugar syrup, which is simply sugar dissolved in water, and the bees will quickly store this as though it were honey.

DISEASES

Bees are relatively free from disease, perhaps because they are always so busy. However, during your weekly summer inspections there are two things you should look for.

Foul brood

This is a disease in which the grubs in the brood chamber die. It is serious and notifiable, and the only course open to you is to kill the bees and burn the hive.

Acarine disease

This, which is sometimes called Isle of Wight disease, is also serious but fortunately can be cured. It is caused by a mite which lodges in the bee's throat and prevents it from flying.

I hope that after reading this chapter you will feel that bees are worth a try. You will find, after you have overcome any initial diffidence about handling so many potential stingers, that they require little attention and, if you manage them well and are reasonably lucky with the weather, they can be profitable. The ideal, of course, is not one hive but several, and this is something you should aim for as you gain experience.

12   The stew pond

You will probably have seen the picture called, as I remember, 'Tomorrow will be Friday'. Some jolly looking monks are sitting round a pond armed with fishing lines and nets, angling for tomorrow's dinner. We had a reproduction on the sitting-room wall when I was a boy, and I remember thinking that this seemed a rather sensible way of keeping one's food.

Stew ponds they were called, ponds stocked with fish which would breed and multiply and keep the abbot's table well supplied. Most manor houses had one, and many of the villagers, if they had a stream nearby, followed their lord's example. If the miller did not keep a stew pond as such he usually did quite well from the pond below his mill.

Now, the question is, can you have one?

Well, obviously you will need natural water, a pond or a stream that can be dammed, and since I have more than a mile (1·6 km) of rippling water running through my own land let me tell you how to go about it.

In the first place remember that the River Board will have certain rights over the stream itself and you won't be able to alter its size or direction without their permission.

What you should do is look for a sizeable area of lower ground adjoining the stream, possibly an area which is already muddy and half filled with water after torrential rain. Get to work with your shovel or, better still and far less back-breaking, with a bulldozer or a tractor fitted with a scoop which you may be able to borrow from a sympathetic neighbour. (Perhaps you can tempt him with the thought of all those fish suppers.)

Make sure that the bank which separates the pond from your stream is higher than both and that it will still act as a safety barrier in mid-winter or in periods of heavy rain.

Now all you want are two lengths of 6-in (15-cm) piping, one to divert water from the stream and into your pond, the other to take the overflow back 20 to 30 yds (18 to 27 m) downstream. Across the mouths of both pipes you will need a small weld mesh to keep the fish from straying.

If possible you should site your pond near a big tree, the exposed roots of which will provide excellent cover for the fish.

Don't sink your pond where there is a risk of pollution from farm buildings – your own or your neighbour's – and keep clear of any sheep dips as the disinfectant will almost certainly kill the fish.

Clean the end of the exit pipe from time to time to avoid any risk of overflow. A friend of mine who failed to take this precaution had ducks swimming up to his back door and a couple of hundred wet and frightened hens huddling on the shed roofs in his yard.

Before you fill the pond you should plant weeds and pond grasses in the bottom and round the sides, for this not only provides a natural cover for the fish but gives them the oxygen they need. Incidentally, undesirable insects such as water fleas breed in and around these patches of weed so be ready with a spray gun.

Suitable water plants are rushes, water marigold, water crow's foot, which has rather pretty white flowers with yellow centres, and water cress. Incidentally, quite apart from the fish pond, you should plant some water cress in the stream if there is none there already as this provides another valuable source of food for your own table.

Now, after a few days, if the pond is still there and shows no sign of running low or overflowing you can think of fish.

The golden rule here, at least for the beginner, is to err on the side of caution, that is, to order less than you think you can manage rather than more. They will soon multiply.

Go to a trout hatchery, and if you have never seen one before you will marvel at the scientific know-how that goes into professional fish-breeding.

When trout hatch from the eggs they are about 1 in (2·5 cm)

long and are known as fry. From fry they grow into fingerlings. If you buy fingerlings and feed them properly they should be 5 or 6 ins (13 to 15 cm) long at the end of a year. Many of the old type fish-breeders have their own special foods by which they swear, but the various fish foods and pellets which you can buy in any pet shop contain all the necessary ingredients.

Recently I saw a first-class pond such as I am describing in a wild life garden at Glyn Neath in South Wales. Idris Hale, the owner, had everything perfectly arranged and had even installed slot machines from which trout pellets could be purchased and fed to his magnificent collection of brown trout. Some of these must have been 18 ins (45 cm) in length and several pounds in weight. If you could breed some like that in your stew pond, and there's no reason why you shouldn't, you would be well off indeed.

Of course, it's not always plain sailing. A friend of mine who had a trout pond at his farm in the Black Mountains was equally successful but was unfortunately plagued by marauders. This is a wild and beautiful part of the country, and it may be that the hazards he encountered would not trouble you. But as a cautionary tale I think it worth recounting.

The first time I found out that he had a stew pond was when I went to stay with him for a few days intending to buy some of his prize Friesians. It was late spring, I remember, the hedgerows were in leaf and the apple trees, which stood like old, green coated soldiers on parade, were heavy with blossom. We had been up to the hill pastures to see his sheep, and, later, we inspected the pigs, a dozen sties of splendid Tamworths, and the Friesians, which were the object of my visit. It was only when he guided me artlessly to the meadows below the farm that I saw the pond.

'What's that?'

'What?'

'The pond.'

'The stew pond,' he said. 'I don't suppose you've seen it before.'

'Of course not. It's new, at least it wasn't there when I came a couple of months back.

My friend shrugged. 'It's just something I thought of, a hobby really. I hope it will pay its way.'

'But what's it for? You've no ducks – or have you?'

'Certainly not. You don't have ducks on a stew pond,' he said, severely. 'There'd be no point.'

I felt, and probably looked, bewildered. 'A stew pond? You mean for fish?'

'What else?'

When we inspected the pond, which he was obviously keen for me to do, I found that it had been formed exactly as I described above, with a tractor to scoop out the earth, 6-in (15-cm) piping to ensure the flow of water and a solid old tree, an oak, growing alongside.

I'm afraid I didn't know much about stew ponds. 'How do you get the fish?' I asked. 'Do they come down the river?'

'Certainly not.' He looked at me pityingly. 'It's stocked with trout, or rather fingerlings. Next year I'll be having them for supper every night.'

This seemed to me to be overdoing it but I said nothing, for to tell the truth I was impressed. A stew pond of your own, a pond where fish could live and breed and multiply: it sounded too good to be true.

And so it proved.

It wasn't until midsummer that I called again, and this time my friend didn't seem so anxious to guide me to the stream below the farm. We saw the sheep and the pigs and the Friesians and some guinea fowl he had just brought from market. 'Better than watchdogs,' he said. 'No one would get within half a mile of the farm without their warning.'

I said, 'I know. I've kept guinea fowl myself. But what about the stew pond?'

He stopped in his tracks and looked distinctly uncomfortable. He mumbled something about not doing too well.

'What do you mean, not doing well?' I asked. 'What happened?'

'Just bad luck,' he said. 'Could have happened to anyone.'

'But what happened?' I persisted. 'I must know, because I've been thinking of starting a stew pond myself.'

He gave me a sardonic smile. 'Why not?' he said. 'You

won't be troubled by otters in your part of the country, let alone mink.'

I said, 'I don't know about that. There are otters in the river, although I doubt whether they come into our stream.'

'Exactly!' he said. 'Soon after you came here last time I found that instead of multiplying the fish were decreasing. When I went to feed them there weren't enough to eat the pellets.'

'You think it was an otter?'

'I *know* it was an otter. I went down there soon after dawn one morning and saw it coming out of the pond with a fish still flapping in its mouth.'

I made sympathetic noises, which were sincere enough for I knew what it was like to lose precious stock. 'So what happened?'

'I got the local hunt out. The hounds followed it upstream for a couple of miles and then lost it, but it did the trick. The otter didn't come back here again.'

I smiled. 'So, all's well that ends well. I suppose you re-stocked?'

'Yes, I re-stocked.'

'And now the stew pond is flourishing?'

'Not exactly.' He pulled a face and then, unable to take his troubles seriously any longer, burst out laughing. 'We lost the otter,' he said, 'but we gained a mink. It wasn't a very profitable exchange.'

'A mink!' I exclaimed. 'Here, in the hills?'

'Oh, yes,' he said. 'The British mink is much more numerous than you imagine. As you know, it's the worst enemy of the fish. Well, we had one in the river.'

'And it found your pond?'

He nodded. 'In two, three nights it cleared the lot.'

'My dear fellow!' I said. 'I really am sorry.'

He grinned. 'Oh, it's not so bad. I still think I was unlucky.'

'What happened to the mink?'

'My stockman shot it.'

'And you got more fish?'

'Yes.' He raised his hand, showing his crossed fingers. 'I hope it will be third time lucky.'

Having got this off his chest he was quite willing, in fact almost eager, for me to re-visit the pond. He took a twelve-bore with him as there was always the chance of rabbits in the lower meadows, and we were walking quite happily down the hill when he stopped.

'For Pete's sake! Look at that.'

I looked and saw and was torn between laughter and commiseration. There, on the edge of the pond, looking rather like a Sunday school teacher I used to know, was a fine adult heron. It had not heard our approach, or if it had it didn't care, for in its beak was a wriggling trout. Even as we watched, it tossed it an inch or two in the air, caught it again and swallowed it in one gulp.

'I'll murder it,' my friend said. 'I'll shoot its blasted head off,' and, suiting action to words, levelled his gun and fired.

The heron looked round, apparently not greatly disturbed, and with a slight hop, launched itself into the air and flapped away.

No doubt it knew as I, and certainly my friend, knew that we weren't going to kill it, for the heron is a protected bird.

Otters, mink and heron: well, if, like my friend, you are unlucky you may be troubled, but I doubt it. Otters are fast disappearing from our rivers, much to the regret of the conservationists. The mink, however, is increasing in numbers and if there is one in your area it will almost certainly pay you a visit. A mink trap baited with chicken's entrails is the answer, and – dare I say it? – you may even be tempted to add this profitable creature to your farming venture.

Herons are another matter. They live and breed in areas of lake and marshland, and if this is your kind of country you may be paid a visit. There is not much you can do about it except to scare them away with a rattle or shot gun when they appear. But don't point the shot gun at them or you will be breaking the law.

DISEASES

Fish are not much troubled by disease, and yours should remain healthy so long as the pond has a flow of running water. Pond weed will give them the oxygen they need, and the vitamins will be contained in the trout pellets you supply.

Only one disease is worth mentioning. If you pick a dead fish with an open wound from your pond, call in the Ministry of Agriculture laboratories at once, for this could be furunculosis. This is a nasty disease which could wipe out the whole of your stock.

Apart from this, and possible losses to predators, your stew pond should flourish. You won't make a fortune, of course, but with luck and the minimum of attention you can have fish suppers whenever you feel like it, which, considering the price of fish today, is something.

If you had reservations about pigs, and perhaps even more about sheep, you will probably think that a milking cow is beyond the scope of your small-holding. You will have read or heard that it is a sensitive creature which needs careful handling at all times, that it will only give of its best if treated with understanding and that the cowman on any good-sized farm is a knowledgeable and experienced specialist.

Well, all this is true, but that doesn't mean that with patience and understanding and a certain basic knowledge you cannot keep a cow successfully: in fact I would go further and say that, having taken the plunge, you could well be surprised at the speed with which you and your new acquisition find rapport.

I have kept cows for as long as I can remember, so I am probably biased, but for me the idea of a pasture without at least one milking cow is unthinkable.

My present favourite is a quiet little Jersey named Buttercup, who is delightfully tame and docile, so much so that I can never walk across her field without her pushing beside me. She gives me 500 to 600 galls (2270 to 2730 l) of rich, creamy milk a year, which is fairly modest compared with the output of some Friesians, but it is more than enough for me. From this we produce as much cream, butter and cheese as we require.

I mention Buttercup because I think she is exactly the sort of cow for you. As you probably know, cows are kept either for beef or for milk or, in a few dual-purpose breeds, for both. The Jersey, from my experience, is one of the best of the milkers and, with the Guernsey, which is rather bigger and therefore not quite so well suited to your backyard farm, shares top place for rich butter fat content.

Since you will clearly be looking for a milking cow I think in fairness I should mention some others.

The British Friesian is the familiar black and white breed you see in most parts of Britain. It is considerably larger than the Jersey, and although its milk yield is unsurpassed (there are several recorded yields of more than 2000 galls (9092 l), the butter fat content is no more than average.

The Ayrshire, as its name implies, comes from Scotland. It is found in many parts of England and in other parts of the world. It is red and white in colour and has a neat dairy-like appearance.

The Kerry is an Irish breed particularly suitable for pasturing on wild, windswept moorlands, since it is exceptionally hardy.

You will have gathered already that if I were in your position with a small-holding of 2 to 3 acres (0·8 to 1·2 hectares) I would plump unhesitatingly for a Jersey. She is on the small side, generally quiet and reliable and exactly suited to your needs.

When buying look for a small, bright-eyed cow with a sleek fawn coat and a friendly disposition. Check, whether you are buying a Jersey or any other breed, that she has been hand milked, for one that has been accustomed to machine milking may react violently and kick you and the bucket into the next stall the first time you try the familiarity of hand milking. It is also important to check that she is accredited i.e. brucellosis free.

PASTURE

But before you rush off to market or to a neighbouring farm you must, in fairness to yourself as well as the cow, decide whether you have sufficient pasture to keep her.

Ideally you should have 2 acres (0·8 hectare of grassland, although you could probably get by with less and would not do much better with more.

Divide your pasture into four half-acre (0·2-hectare) plots, either with a wire cattle fence or, just as effectively, with a single strand of electrified wire.

Perhaps I should explain at this stage that the electric fence which we have mentioned in earlier chapters is quite harmless to animals although it is one of the most effective ways of enclosing a particular plot. The single strand wire, secured to terminals on wooden posts, is connected to a battery which sends out a mild electric current. The animal rubs against it, gets a shock, and is careful not to repeat the experiment. Dry batteries last about two to three months.

Your half-acre (0·2-hectare) plots should be used alternately for pasturing and should see your cow through the spring and early summer. One or two paddocks may provide a crop of hay.

A cow, like other grazing animals, appreciates a change of grass and will increase her milk yield considerably if she is moved backwards and forwards between the plots. Put her into a bigger plot and she will simply fill her belly and then lie on or trample the rest of the grass and spoil it.

The other two plots you should lay up for hay having first given them a dressing of manure. After one early cut of hay and a good application of nitrogen you can usually take a second, although this will not be as heavy a crop as the first. Hay, remember, is your insurance against bad winters and rising food costs, and if stored in a dry, open-fronted shed will keep for a couple of years. (Incidentally, the first dressing of manure can come from your own cow shed.)

FEEDING

But, still before you buy your animal, consider the cost of feeding.

As I said, a couple of half-acre (0·2-hectare) plots of good grassland will provide enough food during the spring and early summer. Later in the year and throughout the winter you will have to supplement what little she can take from the pasture with a diet of hay, root crops and concentrates.

The amount of food required each day depends partly on the time of year and partly on the milk yield. Thus your little Jersey, producing perhaps 1½ galls (6·8 l) of milk a day in the autumn, while she is still out at grass, will need about 5 lbs (2 kg) of hay, 25 lbs (11 kg) of greenstuff such as kale, which is a crop you could easily grow on your vegetable plot, and 3 lbs (1·3 kg) of concentrates. In mid-winter when she is producing, say, 2 galls (9 l) of milk a day she will need about 10 lbs (4·5 kg) of hay, 30 to 40 lbs (13·5 to 18 kg) of swedes or other root crops such as mangolds and as much as 9 lbs (4 kg) of concentrates.

A rough costing based on prices current at the time of writing will tell you that in mid-winter, with concentrates at 5 pence a pound and milk fetching 40 pence a gallon, there is not much margin for profit, but if you accept that self-sufficiency rather than profit is the aim I hope you will still decide that a milking cow is a 'must'.

You should feed her twice a day and, most important of all, make sure that she has plenty of water.

To judge the probable cost remember that you can produce the hay from your own plot and, if you have a vegetable garden, most of the greenstuff and root crops. Your only positive outlay will be for the concentrates, which you can buy in the form of dairy nuts from your local corn chandler. These give a balanced diet containing protein, minerals and trace elements, but, for a change, your cow will enjoy a top-up meal of crushed oats or flaked maize.

HOUSING

Having seen on local farms elaborate and costly cow sheds, with their cement floors and stalls and drains, you are probably wondering how much housing is going to cost. Well, for one small well-behaved Jersey not very much.

An existing shed, provided it is waterproof and airy, can be easily adapted to make a suitable cow house or shippen. You will need a rail along one wall to take the end of a cow chain or yoke which will secure her. Make sure that she has plenty of room to lie down and behind her dig a drain or channel 2 ft (60 cm) wide and 6 ins (15 cm) deep to take most of her droppings. With fresh straw to lie on she will stay clean and comfortable.

Behind the drainage channel you will need a concrete path for yourself.

Against the wall and above the tethering rail you should fix a hay rack, a simple construction of wood or metal ribs through which the cow can pull hay as she browses. But remember that like most animals, except possibly the pig, she is a wasteful eater, so you will need a concrete trough below the rack to catch the hay she drops.

The cow shed should be raked and brushed out daily, for, like other farmyard animals, the cow avoids many diseases if she can stay clean. When you go into milk clear off any muck into the dung channel so that you can rest your bucket on a clean surface.

MILKING

Milking is more difficult than it looks but not so difficult that you can't achieve a reasonable proficiency within a week. You milk twice a day and since you will have no early milk lorries to worry about I would advise about eight o'clock in the morning and four o'clock in the afternoon. This should suit you and will also suit the cow so long as you keep to the same hours every day.

It would be idle to pretend that you can learn efficient milking from a book, but it may be helpful if I set down the usual, and best, procedure. After that, only practice can make perfect.

First of all make sure that everything you touch is clean. The cow's hind quarters, which should be clipped from time to time, must be washed, the udders should be bathed with a soft flannel or sponge. The milk can and urn should be thoroughly scrubbed and then scalded or sterilized with steam. Your milking smock and cap should come freshly laundered, while your hands, which are the only part of you which may come into contact with the milk, must be as clean as soap and hot water can make them, and make sure that they are dry before you start milking.

Now, with all things set, you can approach the cow.

If she has been hand milked before she probably won't even notice you except perhaps to turn her head, her mouth still full of hay, and give you a mildly inquiring look. She assumes, you see, that because you are not unlike the cowman she has been used to that you will know all about milking.

Well, if you are slow and gentle there may be no reason to disillusion her.

Set your stool on her right side and take up a position with the milk can between your knees and your left foot just inside her right rear hoof.

Now in front of you as you bend with your right shoulder against her belly, you will see the udder with four teats. You milk these in pairs, the front two, then the back two. Place your hands on her two front teats and then squeeze and pull. Gently, of course, but firmly: you will soon notice from the cow's reaction whether you are being too rough. You will be rewarded by the sight and unmistakable sound of milk jets spurting into the bucket. If it is the first time you have tried milking it is a moment to savour.

After a time the milk jets will come less easily and then dry up. Let her rest for a moment and then squeeze the teats again with thumb and forefinger and repeat the process until there is nothing more. This is called stripping and it is a necessary process for without it the cow would produce less and less milk.

Now, with 1 gall (4·5 l) or so of milk steaming in your can and the cow looking thoughtful but not uneasy you have to extract yourself and the stool and the can without any mishaps. Remember that there is a dung trench behind you and that the cow, relieved that all is over, will probably choose this moment to move.

Your next job is to cool the milk and this should present no problems; you simply place it in the refrigerator or stand your bucket in a sink full of cold water until the milk is 'cooled' down to 10° C (50° F).

When you drink your first glass of milk from your own cow and, even more, when you eat your own cheese and spread your own butter, you will feel that you have conquered the world. All the hard work and the doubts will be forgotten: risking so much to become a backyard farmer will have been worth while.

BREEDING

It will not be long, I expect, before you think how nice it would be to have not one cow, but two. Your little Jersey will encourage this thinking when she indicates that she is ready to mate.

This is called 'bulling'. As a rule her milk yield will decrease and she will bellow uncharacteristically. As she approaches the oestrum her vulva will swell and you may notice some discharge coming through.

Since it would be clearly impracticable for you to keep a bull you will have to do what other small farmers do, that is rely on artificial insemination. As soon as you see the bulling signs ring up the breeding centre of the Milk Marketing Board, who will arrange for the inseminator to call. Since he will be making a round of calls at several farms in the area try to telephone as early in the morning as possible.

Gestation is just under $9\frac{1}{2}$ months (280 to 284 days). A few weeks before she is due to calve her udder, which was previously loose and floppy, will become firm and swollen. A good stockman always increases the feed to a cow in calf for several weeks prior to calving.

Care of the calf

Some cowmen take the calf from its mother immediately after birth on the principle that she will not fret for what she has never known. I am afraid that I disagree with this, mainly because I think that the calf needs its mother and the thick yellowish milk, called colostrum, she provides, especially during the first critical days. The calf should clear this colostrum in about three or four days, and this is the more usual time for cow and calf to be separated.

Directly after birth the cow will dry the calf by licking. Then is the time to give her a refreshing drink. For the first two or three days she will only require a light diet, but she should then return to her normal feed. Incidentally, if the mother is a heifer, calving for the first time, don't be alarmed if she eats the afterbirth. This is a perfectly natural reaction which is done to restore her body tissue.

If you want to give your calf a really good start it is a good idea to take only half the mother's milk after the first few days and let the calf suckle the rest. A calf that suckles naturally will grow more quickly and be less prone to disease than any calf that is hand reared.

If your calf is a heifer you will want to do everything possible to keep her for she will be a valuable property. If it is a bull calf your best bet will be to keep it for a fortnight and then sell it at the local market. This sounds mercenary, I must admit, but rearing a bull calf will never be a viable proposition for the backyard farmer.

You can, if you wish, hand rear your calf, although, as I said before, it will be more troublesome and prone to disease than if it had been left with its mother.

Soon after it is born you should feed it 2 pts (1 l) of colostrum taken from the mother. This should be warmed and fed from a bucket, a troublesome business at first since the calf instinctively lifts its head to feed and must be persuaded by the judicious use of your fingers in the milk to act as sham teats that there are other ways of feeding.

It will need feeding three times a day for the first week and twice a day, with about 2 quarts (2·5 l) of milk a time, after that.

At the end of the month you can start the calf on a little hay and a few calf pencils. The calf can be weaned when it is eating 2 lbs (0·9 kg) daily – about five to six weeks old.

The weaned calf should receive calf pencils ad-lib until it consumes 4 lbs (1·8 kg) daily and then be rationed to that amount. It can be turned out to grass when about three to four months old.

DISEASES

Some diseases, such as foot and mouth and bloat, which also affect other animals, have been described in earlier chapters. The main diseases peculiar to cattle are as follows.

Milk fever

This is due to a sudden loss of blood calcium immediately before and after giving birth. Usually within one to two days after calving the infected cow begins swaying and becomes unsteady on her legs. She will sink to the ground and after several vain attempts to rise will lie quiescent until she sinks into a coma.

If you see these symptoms act without delay. Put some bales of straw around her to prevent her falling and send for the vet. He will give her one, or possibly two, calcium injections which will relieve her distress and, if you have acted quickly, should ensure her recovery, and she will get on her feet very soon.

Herefordshire disease or grass tetany

This is a disease associated with a fall in the blood magnesium. However, it is not connected with calving, but is generally attributed to a sudden change of diet such as a complete change from winter feeding to grazing.

In severe cases the cow has convulsions, which are fatal: more usually it becomes nervous and develops a curious stumbling gait.

The less severe cases can sometimes be cured by injections of calcium and magnesium salts.

Mastitis

Inflammation of the udder, called mastitis, can be due to several causes, but the most usual type is a chronic condition, first noted when milking. One of the quarters will be hard and hot, and you will have difficulty in getting any milk through. What little you get may be flecked with blood and will appear in clots.

Fortunately this chronic type responds well to veterinary treatment.

However, the milk from this quarter is infectious and should be collected separately in a tin and disposed of down the drain – *not* anywhere in the cowshed. Milk from other quarters should be healthy.

14 Self-sufficiency

As we agreed at the beginning of this book, backyard farming is an exercise in self-sufficiency. Meat and eggs, butter, cheese, honey, milk and fish: all these could come from your plot. You could stand aside, as it were, from the spiralling cost of living and by your own hard work and enthusiasm build a life of your own, free of rising prices and the imported foods which cost our country dear. You would also be striking a blow for Britain.

I wonder if you have ever flown across this country of ours, from Bristol to Glasgow, say, which was one of my regular journeys when I was broadcasting. If it's a clear day and you fly below the clouds you will see, from 15000 ft (4500 m), vast stretches of countryside. It is a marvellous sight, the cities appearing as mere clusters of buildings huddling beside rivers, the towns no bigger than villages, and the villages scarcely recognizable except as the focal points of farm tracks.

But what will impress you, I am sure, is the vast greenness of it all. A few stretches of brown where the land, perhaps recently tilled, is not yet fresh with corn, a fair amount of heathland spiked by outcroppings of rock, the mountain ranges of Snowdonia, and that's about all: everything outside the towns is green, green pastures, green marshland, green forests, green meadows and – you will even see this from 15000 ft (4500 m) – green backyards. If you think about it I am sure that you will feel as I do that in these harsh economic times there are still vast areas of land which are unproductive.

Well, in your half-acre (0·2-hectare) backyard or your small-holding you can do something about it. When I started this book I had no intention of beating any patriotic drums and really I am only doing so now for good measure. Back-yard farming is about self-help, a common-sense solution to your own economic problems: it is rather gratifying to know that as you help yourself you will also be helping the country.

More than once in this book I have talked about sharing your problems with neighbouring small-holders and farmers. You will find this easier than it sounds. The farming community, like any other group of experts, likes to keep its expertise to itself. This is only natural. But farmers by tradition or perhaps by the nature of their jobs are steady, philosophical people, and I have never found one yet who was not willing to help a newcomer. When you think of it perhaps that's not surprising. To one who spends the best part of his life in quiet fields and hay-scented barns dealing with the age-old problems of cultivation and animal care the advent of a newcomer and possible competitor, although on such a small scale is hardly a cause of resentment.

The best place to meet them is at market or at farm sales where, as like as not, they will talk to you without any need for introduction, for – and I speak from experience – no farmer, however experienced, can smother certain doubts when he comes to invest large sums of money in livestock. So he will talk, not so much for the advice you might give as for the relief in discussing his doubts aloud. Lend a sympathetic ear and I would be surprised if in a very short time you were not friendly with most of the farmers and small-holders in your area.

I remember some few years back visiting a friend in Wales. He had a hill farm near Glaenau, 500 acres (200 hectares) of moorland pasture, more heather than grass and enough out-lying rock to build a castle. Life wasn't easy for him, although I never heard him complain and indeed he seemed to make a decent enough living.

When I arrived at the end of a bumpy mountain track his wife, who met me in the doorway, said that he was out. That was not unexpected, for in this, the lambing season, he spent most of his time, day and night, up the mountain.

But it wasn't that. 'It's Williams *bach*, down in the village,' his wife said, 'having trouble with his calf.'

'Williams the seedsman?' I asked. 'How long has he been farming?'

Not long, it seemed, and that was the trouble. Williams was a pleasant, fresh-faced lad, a child of the village, who had left school at fifteen to take a job as travelling salesman for a firm of corn and seed merchants in Caernarvon. A good job, everyone said, with regular hours and none of the risks of farming.

But it wasn't good enough for Williams, who had been born and brought up among farmers and had been used to farming talk since he was a child. No, like any other true farmer, he had to have stock, and this calf, bought from his meagre savings, was his first venture.

When I walked down to the village I had no difficulty in finding the Williams cottage. Half a dozen farm carts were lined up there, next to a plough horse and a couple of Land Rovers. In a small shed at the back Williams and his friends were tending the sick calf.

No one noticed me enter, so intent were they on the wretched animal, and I was able to recognize from its heaving flanks and saliva-flecked mouth the tell-tale symptoms of calf diphtheria.

'No chance; you'll be wasting your time,' most farmers would say, knowing that the mortality from this disease is high.

But not the farmers of North Wales. They rallied round, one had already driven off to intercept the vet, and there they stood, forgetful of their own problems, the ewes that were lambing on the mountain or the cows calving in the byres. Williams *bach*, a newcomer to their community, although no stranger, was in trouble and they were determined to help.

I wish I could give you a happy ending to this story, but in fact, despite the efforts of all those well-wishers, the calf died.

But there is a moral to this tale, as I am sure you will see. When the farming community accepts you into its ranks you are treated as one of their own. The local farmers and small-holders should be your friends, and will be if you give them half a chance, and their knowledge and good will can be your insurance against failure. The successful farmer in your area, backed by experience and knowledge of local conditions, can give you more useful information than you can get from this or any other book, and you will find it information freely given. In Devon they say if you want to learn farming, you look over a good farmer's fence and watch what he *does*.

This is a textbook, slightly irreverent, I'm afraid, but still a book written to inform. If it encourages even a few readers to branch out from humdrum lives to the greater joys of farming it will have achieved its object. It will, I hope, give you all the basic information you require and tell you where to go and what to do when you come up against problems.

As I write I think of you in one or two years' time, or perhaps sooner, waking up at dawn and going downstairs to scour the milking bucket. Very soon your wife will follow, and as she brews a pot of tea she will be warming mash for the hens, bran for the rabbits and meal for the pigs.

In the larder will be a plump turkey, the remains of yesterday's dinner, pots of honey and enough butter, milk and cheese to feed a regiment. In the deep freeze will be sides of bacon, geese, rabbits, guinea fowl and legs of lamb. Last evening on the radio they were talking about yet another rise in the cost of living and about ordinary families who could no longer afford a week-end joint.

But not you. It will be difficult, I expect, to avoid a feeling of complacency, even smugness, but in defence you will be able to say, truthfully, that all you own has come from courage and hard work

The backyard farmer is, in my opinion, the man of the future, his contribution to the economy will be increasingly important: he is one of the few, the very few, who will be able to watch the rising cost of living without regret.

It is a sobering thought and quite a responsibility, but one which you will enjoy.

Go to it then and enjoy the good life. I wish you luck.

Approximate conversion tables
– ready reckoner

| kilogram to lb | | lb to kilogram | | kilogram to cwt | | cwt to kilogram | |
kg	lb	lb	kg	kg	cwt	cwt	kg
0·5 =	1·10	0·5 =	0·23	25 =	0·49	½ =	25·4
1 =	2·20	1 =	0·45	50 =	0·98	1 =	51
2 =	4·41	2 =	0·9	100 =	1·97	2 =	101
3 =	6·61	3 =	1·36	150 =	2·95	3 =	152
4 =	8·82	4 =	1·8	200 =	3·94	4 =	202
5 =	11·00	5 =	2·26	250 =	4·92	5 =	254
6 =	13·20	6 =	2·7	300 =	5·90	6 =	305
7 =	15·40	7 =	3·1	350 =	6·89	7 =	355
8 =	17·60	8 =	3·6	400 =	7·87	8 =	404
9 =	19·80	9 =	4·0	450 =	8·86	9 =	456
10 =	22·00	10 =	4·5	500 =	9·84	10 =	508

| Litres to gallons | | | Gallons to litres | | |
litre		gallon	gallon		litre
0·5	=	0·11	0·5	=	2·27
1	=	0·22	1	=	4·55
2	=	0·44	2	=	9·09
3	=	0·66	3	=	13·64
4	=	0·88	4	=	18·18
5	=	1·10	5	=	22·73
10	=	2·20	10	=	45·46

AREA

Hectares to acres			Acres to hectares		
ha		acres	acres		ha
0·5	=	1·24	0·5	=	0·2
1	=	2·47	1	=	0·4
2	=	4·94	2	=	0·8
3	=	7·41	3	=	1·2
4	=	9·88	4	=	1·6
5	=	12·36	5	=	2·0
10	=	24·71	10	=	4·0

Square metres to square feet			Square feet to square metres		
m^2		ft^2	ft^2		m^2
1	=	10·76	1	=	0·09
2	=	21·53	2	=	0·18
3	=	32·29	3	=	0·27
4	=	43·06	4	=	0·36
5	=	53·82	5	=	0·46
10	=	107·62	10	=	0·92

LENGTH

Millimetres to inches			Inches to millimetres		
mm		inch	inch		mm
25	=	0·99	1	=	25·4
50	=	1·97	2	=	50·8
100	=	3·94	3	=	76·2
200	=	7·87	4	=	102·0
300	=	11·80	5	=	127·0
400	=	15·70	6	=	152·0
500	=	19·70	12	=	305·0

Metres to yards			Yards to metres		
m		yard	yard		m
0·5	=	0·55	0·5	=	0·46
1	=	1·09	1	=	0·91
2	=	2·19	2	=	1·8
3	=	3·28	3	=	2·7
4	=	4·37	4	=	3·6
5	=	5·47	5	=	4·5
10	=	10·90	10	=	9·1

Useful addresses

The Ministry of Agriculture, Fisheries and Food, Whitehall Place, London SW1

Department of Agriculture and Fisheries for Scotland, St Andrews House, Edinburgh 1

The Agricultural Research Council, 160 Great Portland Street, London WIN 6DT

Agricultural Development and Advisory Service, Ministry of Agriculture, Fisheries and Food, Great Westminster House, Horseferry Road, London SW1

National Agricultural Centre, Stoneleigh, Kenilworth, Warwickshire

Land Settlement Association, 43 Cromwell Road, London SW7 2EE

National Farmers Union, Agriculture House, Knightsbridge, London SW1

The Meat and Livestock Commission, PO Box 44, Queensway House, Bletchley, Bucks.

The Poultry Club, 72 Springfield, Dunmow, Essex

British Waterfowl Association, Belle House, 111/113 Lambeth Road, London SE1

Eggs Authority, Union House, Eldridge Road, Tunbridge Wells, Kent

National Pig Breeders Association, 7 Rickmansworth Road, Watford, Herts. WD1 7HE

British Landrace Pig Society, Yorkersgate, Malton, Yorkshire

British Goat Society, Palgrave, Diss, Norfolk

National Sheep Association, Tring, Hertfordshire

Ryeland Flock Book Society, 11 Blackfriar Street, Hereford

British Rabbit Council, 302 Farnborough Road, Farnborough, Hants.

British Beekeepers Association, Rides, Warden Road, Eastchurch, Sheerness, Kent

British Cattle Breeders Club, Lavenders, Isfield, nr Uckfield, East Sussex

Milk Marketing Board, Thames Ditton, Surrey KT7 0EL

Dexter Cattle Society, Lomand, Seckington Lane, Newton Regis, Tamworth, Staffs. B79 0ND

The British Friesian Cattle Society, Scotsbridge House, Rickmansworth, Herts.

The British Guernsey Society, Giggs Hill Green, Thames Ditton, Surrey

The Hereford Herd Book Society, Hereford House, 3 Offa Street, Hereford

Jersey Cattle Society of UK, 154 Castle Hill, Reading, Berkshire

Index